Alastair McIntosh is an in[...] and honorary professor at C[...] A human ecologist, he speaks widely on the challenging questions of our time including climate change, globalisation, land reform, community empowerment and nonviolence – with an emphasis on psychological and spiritual depth. He is the author of the bestselling *Soil and Soul: People versus Corporate Power* (Aurum), described by George Monbiot as 'a world-changing book', and *Poacher's Pilgrimage: An Island Journey* (Birlinn), reviewed in the *Times Literary Supplement* as 'fascinating, provocative and, occasionally, very funny'.

RIDERS
ON THE
STORM

THE CLIMATE CRISIS
AND THE SURVIVAL
OF BEING

ALASTAIR
McINTOSH

BIRLINN

First published in 2020 by
Birlinn Limited
West Newington House
10 Newington Road
Edinburgh
EH9 1QS

www.birlinn.co.uk

ISBN 978 1 78027 639 7

British Library Cataloguing in Publication Data
A catalogue record for this book is available from the British Library
Typeset by Initial Typesetting Services, Edinburgh

Printed and bound by Clays Ltd, Elcograf S.p.A.

To my dear mother, Jean Patricia McIntosh,
who when I was a child encouraged me
to write stories.

CONTENTS

INTRODUCTION

Not fare well,
But fare forward, voyagers.

T. S. Eliot

'Into this house we're born,' as Jim Morrison's strange prophetic ballad had it. In part we lack a choice because 'into this world we're thrown'. Climate change is simply where we're at. It is where the evolution of conscious life on earth has brought the planet to.

But we can make choices as to where we go now. Both individually and collectively we can choose to evolve culturally. It is with the dignity of life on earth, and our human part in it, that the passion of this book is concerned.

I wrote it in the Scottish city of Glasgow, a mile away from where the United Nations hope to hold their climate change summit with world leaders, known as COP 26. Originally, it was planned for November 2020, but the COVID-19 pandemic led to its rescheduling to November 2021.

While there is no obvious way in which the coronavirus is directly linked to climate change, the situation that it precipitated remains enfolded in the greater and emergent planetary crisis caused by greenhouse gas emissions. Lessons learned from one about fragility, and the need to build resilience into our social and economic systems, will transfer over to the other.

My zest in writing this work has been to go beyond the outward science, policy and politics of climate change. Most certainly, I want to summarise and honour those. They are our starting points. But I want to use them as a springboard into the deeper question of *being* itself: a wake-up call, as it were, that quickens to the nature and survival of our deepest humanity.

In the first four chapters, I summarise in plain language, for readers who might need and want it, the current science, context and proposed remedies surrounding global warming. This will stick closely to the peer-reviewed publications of the United Nation's Intergovernmental Panel on Climate Change. If such science and the technological possibilities are not your forte, please feel free to skim over these parts should I fail to entice you in.

In the mid chapters, I explore both the psychology of climate change denial, and alarmism that exceeds the scientific consensus over such debates as near-term human extinction. The contrasting dangers of denial and alarmism are not symmetrical. Denialism has done far more harm than alarmism, and its political drivers differ. However, if we yearn for social justice and environmental sustainability, we must be 'critical friends' towards our own move-ments. I will therefore round on leadership questions in activism in particular. This has implications that go well beyond climate change alone.

From here, I move to looking at what the public sector, the private sector and the voluntary (or wider 'vernacular') sector can contribute in the necessarily rapid move towards a zero-carbon world. This will touch on and critique a range of possible tech-nical interventions such as carbon dioxide removal, the realm of corporate innovation and ethics, the fraught and yet potentially liberating debate around population and consumption levels, and the policies and politics of green new deals. All of a sudden, the latter has found fresh impetus under the ravishes of COVID-19, and with it the pressing need to think through forms of global economic stimulation that don't merely multiply our problems. Here is the opportunity to produce enduring fruit from a new-found public recognition that resilience is not a luxury.

Readers of my other works will know that my approach is far from conventional. Buyer beware! In case such a style is not for you, be warned that I love few things better than moving from hard science to spiritual reflections by a Hebridean sea loch. This is not a how-to book that tries to replicate all the others that do a better job on recycling rubbish, changing light bulbs, or technology and governmental policy options. My interest is to invite my reader on a journey into the survival and thriving of *being* – the being of both human and all other forms of life on earth.

In my closing chapters, I therefore enter further into depth psychology and beyond. To the best of my limited abilities, I examine what it takes to reconnect with the earth, with spiritual life and with one another. With soil, soul and society.

I approach this through a shift into storytelling mode. In a case study, I give an account of going back, in 2019, to my home village on the Isle of Lewis with a delegation of community leaders from West Papua – a province of Indonesia in western New Guinea. The experience shed some astonishing light upon our predicament. It illustrated both the deepest traumatic drivers of the world into which we're thrown, and pointed to some thrilling paths of resolution.

While this is not a book of optimistic platitudes, neither is it counsel for despair. Climate change can press us all to deeper layers of reflection than we might ever have entertained before. Such is our basic call to consciousness. Here might be the freeing up of long-blocked wells, and this for the survival of being in us all. To open up the flows of what gives life.

A crisis is too good a chance to waste. There is a gift, as well as dread, in living through these times. The world on us depends, which begs a question. How can we be riders on the storm?

ALASTAIR MCINTOSH
Govan, Glasgow, 2020

A WALK ALONG THE SHORE

There are places you can go from where the whole world passes by. Little corners, from which to dig from where we stand. Vantage points from which, in a single glance, one can witness some of the key effects of climate change in a landscape, and glimpse the scientific complexity that plays out through space and time. Occasionally in such places, the very stones laid on the ground can tell a story that, as we will see later in this book, can shed a striking light upon geopolitics of our time.

It was April of 2019. I had led a delegation to my home village of Leurbost (roughly pronounced *leu-er-bost*) on the Isle of Lewis, the most northerly of the Outer Hebrides of Scotland. We were some twenty people, including our local hosts. Our skin colours ran full spectrum, from lily white through shades of honey to burnished ebony. These improbable visitors had just arrived from Papua and West Papua, two provinces in the western half of New Guinea that are legally part of Indonesia. My connections with that corner of the world and specifically with neighbouring Papua New Guinea in the east, went back to the 1970s and '80s. Then I had spent four years as a VSO volunteer, teaching down on the coast amongst the Elema people, and setting up small-scale hydro-electricity systems in a couple of the mountain settlements of the Kamea people.

As village leaders, the Papuans had come to Lewis to study land reform trusts and community empowerment. There were reasons why they needed to do this in Scotland. Suffice to say that the visit built on earlier work that my wife, Vérène Nicolas, and I had carried out with civil servants and legislative council members around climate change, land use and collaborative leadership. On this occasion, and working not with officials but with the grassroots, our theme was *Healthy Community, Healthy Land: rediscovering the art of community self-governance*.

Ecology is the study of plant and animal communities. Communities are about relationships. Just as you can have mouse or giraffe ecology, so human ecology studies interactions between the social environment and the natural environment in which we live; you could say, between human nature and natural nature. Anthropogenic climate change – that which has its 'genesis' or origins in the 'anthropos', or human domain – can only be understood in such a framework. That's what the Papuans had come both to explore and to share with our island communities from their own tough-wrought experience.

A Study Tour of Home

Our visit took place on a sunny afternoon. Vérène and I had set it up with 'Rusty', the village blacksmith and chair of the Historical Society. He and I had known each other since our first day at the local primary school in 1960. We gathered at the head of the long arm of the sea that is Loch Leurbost. The wind had dropped down to a breeze. Behind us, the village homesteads stretched out along the road, most of them with narrow strips of arable croft land that ran down to the sea.

We were headed out across the river, over to a rugged stretch of hills inset with gaunt north-facing cliffs that were scraped bare by the last ice age. The heather moors this early in the year were devoid of any hint of floral colour, but one spot stands distinctively apart. In the far sheltered corner where the sharp slope breaks, a bright green sweep of pasture runs on down to the shore. Nestled there upon it are ruins of what we'd call a *clachan*, a cluster of what once had been some four to six small homesteads. These had been constructed in the dry-stone manner, where skill made up for want of mortar.

This whole area was our childhood playground. In the river we'd catch brown trout with our homespun bamboo rods. Off the rocks in early autumn we'd get the 'cuddies', small ocean-going fish that made a tasty fry, and somehow it was always Alex George Morrison who'd catch the codling. Nearby we'd gather shellfish by the gallon bucket to cart home on our handlebars to feed our families, and in the ruins of the clachan, we'd play amongst the tumbled stones and take our shelter from the frequent shrapnel squalls of driving rain. The thatch and rafters of these 'blackhouses', as they're known, had long since settled back into the soil. 'Just something from the old days,' we'd be told, on rare occasions when we'd even think to ask about abandoned places such as this.

We'd timed our visit for the falling tide. That would let our Papuan guests, neither used to icy streams nor equipped with rugged footwear, to ford it at the shallows where it braids before it slips into the sea. So it was that we splashed across and wandered on towards the bright green swathe. Evelyn Coull MacLeod from the village had brought a huge enamel yellow pot that was filled with kindling. She carried it by one handle with her friend, Catherine Mary Maclean, on the other. As we walked, I pointed out to everyone the undulations in the land. Every three or four metres the ground rose and fell in ridge and furrow, and streaked down the hillside to the shore in long and curving parallel rows.

These are the *feannagan* – the *fee-an-a-gan* or 'lazy beds' of raised-bed agriculture. Once they would have bristled come the summer with crops of potato, barley and oats. Their soil was made by sweat and brawn. Heaped up in ridges, the infertile peat gave both depth and drainage. Each spring, it was enriched with animal dung and rotted seaweed for fertility. The roofs of old were without chimneys. Smoke percolated through the thatch. When that required replacing, it too was dug into the land. So it was that, wasting nothing, even nutrients in the soot were captured and recycled. When tin roofs came in, some folks saw them as a mark of poverty. They were so cold before the days of insulation. I can remember an old man just across the road from where we'd walked down with the Papuans complaining to me of that cold. They were noisy, too, when it rained, especially in the hailstorms. And how was a tin roof going to give fertility back to the land?

Such was a living human ecology, a remnant of those intimate relationships between the human environment and the natural environment that much of modern life seems almost to have made redundant. Here was a system of subsistence agriculture with closed nutrient loops that had endured sustainably for hundreds, and in some parts of the island, even thousands of years. It was the old indigenous 'permaculture', no need here to have such notions imported from Australia. As so often in ecology, our way works best for our place.

On its own, peat is soft and mushy, and the underlying layer of gritty boulder clay is thick and hard to work. The bedrock of ancient, crystalline Lewisian gneiss gives grudgingly of its few nutrients. And yet, these rimples of *feannagan* sport quite the loveliest of soils. Fine and crumbly, as if a dark volcanic loam enriched with humus, they're turbo-charged with potash and trace elements. They're also high in lime from seashells that came with the basketfuls of seaweed. Even into my childhood, the food that they produced augmented diets of dairy, meat and fish for many families. At primary school, what elsewhere calls half-term to us was the October 'potato holiday'. No holiday it was for grassroots families. You'd see even the youngest children helping out their parents, digging up the crop by hand and putting it in winter store – the big ones for the people, the small to feed the cow and hens.

While such work was back-breaking, just as pounding sago palm would be for the Papuans, ample testimony speaks to an essential wholesomeness of these ecologically attuned ways of life. In the 1930s, Dr Weston Price of the American Dental Association conducted surveys to compare indigenous communities around the world. He reported on 'the superb health of the people living in the Islands of the Outer Hebrides . . . characterized by excellent teeth and well formed faces and dental arches.' In one place that had no shop to import modern foods, he found just one cavity per hundred teeth examined, and of the isles in general, formed this conclusion: 'Life is full of meaning for characters that are developed to accept as everyday routine raging seas and piercing blizzards representing the accumulated fury of the treacherous north Atlantic. One marvels at their gentleness, refinement and sweetness of character.'[1]

Reading an Eroding Landscape

I am certainly not going to suggest that we all go back to black-houses for a bygone ecotopia. But in coming here I would ask you to make a disturbing observation. There ought to be little if any natural erosion at the head of a long, narrow and sheltered fjord like Loch Leurbost, that has a stream depositing fresh sediments. But these days, the ends of the *feannagan* are sharply truncated. Bare earth spills from each ridge, as if a row of little dumper trucks are tipping their spoils away into the sea. The coast looks bitten, as if in some dystopian dream sent to confute the praise of Dr Price a giant set of rotting teeth had taken an avenging bite. Each full tide nibbles at the rich black soil. It hurts me knowing what comprised the making. In places where it's all gone, one sees a sight most unexpected. Resting on the clay laid down at the termination of the last ice age, is a ghostly forest of ancient tree stumps.

Once, stands of pine had covered much of the island; indeed, much of the far north of Europe. Typically, bog wood from locations such as this dates back around 5,000 years. It seems that the decline was partly caused by human impact, but as the scientists say, 'most likely triggered by climate change'.[2] A relatively rapid change to wetter weather and waterlogged soils caused the pines to struggle on flat or lower ground. It favoured sphagnum moss with shrubs and heather. These built up in time as peat, that in its turn engulfed and then preserved the remnants of the trees. If their wood is dried and cut, it still burns brightly on the fire, and with a resinous aroma as good as any fancy-fangled bathroom spray.

How does one read such features in a landscape? What brings the shifts that makes an ancient forest first to die, and then to see its remnants start to slip beneath the waves? At Leurbost, one can even see the intricate patterns of the finer roots. They weave a floral knotwork on the sheets of mud exposed at just below the high tide mark. As for the *feannagan*, they're like barometers of latter-day erosion, black earth in place of silver mercury. What factors are at play here? What of the climate, past, present and future, might such sites teach? And as we will examine later, what might the human ecology that cries witness from this place teach about the politics of climate change today? First, to get our bearings, we must set out some principles of science and measurement.

Greenhouse Effect and Carbon

The earliest and simplest forms of life began within a billion years of the earth's formation some 4.5 billion years ago. It was not for another 4 billion years – bringing us to just over 500 million years before present – that advanced life burst into a dizzying array of evolution in the so-called Cambrian Explosion. Throughout that time, what had been a toxic atmosphere was transformed by life itself. From the most primitive microscopic species upwards, plants and animals captured and stored carbon in limestones, coal, oil and gas strata, and in the living skin of forests and their soils. James Lovelock called this process Gaia, named after a Greek goddess of the earth. When we burn these to release the pent-up energy of ancient sunshine, or when we heat limestone to make cement (which accounts for up to 8 per cent of global carbon dioxide emissions),[3] we are unravelling aeons of the planet's housekeeping work. Carbon gases have a molecular structure that trap heat. Acting together with water vapour, they serve like a blanket wrapped around the earth that prevents the sun's heat from escaping back out to space.

This is the so-called greenhouse effect. It was long assumed that the first person to confirm it experimentally was John Tyndall at London's Royal Institution in 1859. However, in 2010 the earlier work of an American inventor and women's rights campaigner, Eunice Newton Foote, was by chance rediscovered. She found that a jar filled with carbon dioxide, left in the sunshine with a thermometer inside, became very much hotter and held the heat for very much longer than one filled with ordinary air. Her paper was read on her behalf in 1856 by a senior male professor at the American Association for the Advancement of Science. She clearly saw the implications that accounted for the earth's atmospheric temperature, and this was reported at the time in *Scientific American*. Thereafter, until her recent rediscovery, the mother of global warming science was forgotten.[4]

The year 1750 is often taken as a rough-and-ready date to mark the start of the Industrial Revolution, but 1850 is now used as the benchmark from which to distinguish between pre-industrial and later temperatures as this, still quite early in the era of industrial emissions, was when more reliable instrumental measurements became available.[5] In our present, evolved form as *Homo sapiens*, human beings have roamed the planet for only about 300,000 years, and ice cores show that throughout this time, and prior to the twentieth century, CO_2 levels stayed well below 300 parts per million (ppm).[6] As of January 2020 and as recorded at the global monitoring observatory at Mauna Loa in Hawaii, it stood at 414 ppm.

Those who deny climate science sometimes say that that's a tiny amount. It's only 0.04 per cent of the atmosphere, less than a twentieth of 1 per cent. How could that do harm? Well, picture it like this. Scotland's alcohol limit for driving is 50 milligrams per 100 millilitres of blood. As alcohol has a lower density than blood, that sets the drink-drive limit at just below 0.04 per cent by volume. Our whisky is quite the best, but at 414 ppm, you're banned.

For the first half of the industrial era, the UK as both industrial pioneer and imperial power was the world's largest CO_2 emitter. It was overtaken by the US in 1888, and China took the lead in 2006.[7] Of cumulative anthropogenic (or human-caused) CO_2 in the atmosphere that built up from 1750 to 2017, the USA accounts for 25 per cent, the nations of the EU (including the UK historically) 22 per cent, and China under 13 per cent.[8] Today, based on 2016 figures, China's CO_2 emissions on a 'consumption' basis – attributable to its own citizens – were 14 per cent lower than as measured on a 'production' or 'territorial' basis. The latter counts in emissions from the balance of imports and exports. It is used for international carbon accounting for target setting to avoid double counting between nations. Put simply, the Chinese make our stuff but get blamed for our consumption. In contrast, the UK's consumption-based emissions were 40 per cent higher than its production emissions.[9] We shuffle paperwork while China does the heavy lifting.

On a production basis, the average CO_2 emissions per person in the world were 4.8 metric tons in 2017. China stood at 7.0 tons per capita. Some European countries were surprisingly low, with Sweden at 4.2, France 5.5 and the UK 5.8. Germany was a big hitter at 10.9 because, while exiting nuclear power, it burns a lot of coal. Bigger still was the USA at 16.2, Australia 16.9 and Saudi Arabia 19.3. Qatar topped the league at 49.2, which shows what people can get through if allowed the chance. At the other extreme, Malawi, Chad and Niger have a miniscule 0.1. Kenya has just 0.3 and even oil-rich Nigeria only 0.6. Papua New Guinea stands at 0.9, both India and Indonesia, 1.8, and both Brazil and Egypt, 2.3.[10]

But wait. Is there not something suspicious in those figures? How come the UK seems so low? It depends on what is being counted and how. Add in what the UK imports, and the CO_2 'consumption'-based figure jumps up 40 per cent. Add in the methane mainly from agriculture, nitrous oxide from vehicles and fertilisers, and industrial greenhouse gases such as hydrofluorocarbons used in refrigeration, and then express those as CO_2eq – that is to say, as CO_2 *equivalents* for their potential to force global warming – and the 'true' figure gets a lot more complicated to calculate. The British government's environment department, DEFRA, made a commendably honest attempt to do so in 2019. The uncertainties involved forced it to designate the findings as 'experimental statistics'. It found that the UK's total 'consumption'-based CO_2eq emissions were 784 million tons using 2016 figures. Divide that by the UK population of 65.7 million as it was then, and the consumption-based per capita carbon footprint comes out at 11.9 tons CO_2eq.[11]

To sum up. On the minimalist measure of 'production'-based CO_2 alone, the UK can boast a carbon footprint of only 5.8 tons per capita. On the maximalist measure that is both 'consumption'-based and CO_2eq, it bangs in at 11.9 tons per capita. The difference is a factor of 100 per cent, with a spectrum of positions that lobbyists and politicians can play with in-between. Darrell Huff, who wrote *How to Lie with Statistics* in 1954, will be rejoicing in his grave. For us remaining mortals, God made experts for a reason.

Not Too Hot, Not Too Cold

By the start of the twentieth century, the CO_2 concentration in the atmosphere had exceeded the highest levels of the past 800,000 years.[12] The 300 ppm threshold was crossed in 1910. By the early 1970s it was increasing by 1 ppm per annum. By the start of the new millennium, this had doubled. By 2016 it had exceeded 3 ppm and was still rising relentlessly. The 400 ppm concentration, as measured at the Mauna Loa observatory in Hawaii, was crossed in 2015.[13] Globally, three quarters of greenhouse gas emissions come from fossil fuels and industrial processes such as cement-making. Agricultural practices, such as forest felling, burning and peatland destruction account for most of the balance.[14] It is one thing to make progress on decarbonising the electricity supply. In the UK, for instance, renewables accounted for 29 per cent of generation in 2017, with nuclear at 21 per cent and fossil fuels at 50 per cent.[15] But even in such a renewables-rich country, electricity accounted for only 18 per cent of primary energy consumption by fuel type. The higher-hanging fruit remained the fossil fuels used in transport, industry and heating.[16]

Weight-for-weight, CO_2 accounts for three-quarters of the mix of greenhouse gases in the atmosphere. However, because most of the other greenhouse gases are more potent, this represents only two-thirds of emissions-related global warming since the Industrial Revolution began. Specifically, as calculated in the US government's Annual Greenhouse Gas Index for 2018, CO_2 accounts for 66 per cent of emissions-related global warming, methane for 17 per cent, nitrous oxide 6 per cent, and other mainly industrial gases – a surprisingly considerable 11 per cent.[17] It is the combined effect of all of these that has caused the average temperature of the world to increase by at least 1°C since the pre-industrial era. Two-thirds of that warming has taken place since 1980.[18] A single degree might not sound a lot, but if the human body's temperature is elevated by that amount above its 37°C norm it is considered to have a fever. The planet is less sensitive than the body, yet it is salutary that during the last ice age the global average temperature was only about 5°C below what it is now, and that life thrives within a 'goldilocks zone' of not too hot, not too cold, but just right.

At which point, we can come back round again to those pine stumps at the head of Loch Leurbost. Let me assume that they, like many similar stands, date back around 5,000 years. With the caveat that the science of reconstructing climates after the last ice age is incomplete, what is the history that unfolds before us here?[19]

CO_2, Methane and Prehistoric Climate Change

The chemical composition of air bubbles trapped in ice cores drilled from the Antarctic gives a timeline of the earth's past atmospheric composition. The gradually unfolding picture suggests that anthropogenic climate change may have started considerably further back in history than was once thought. At the end of the last ice age, back to about 12,000 years ago, resurgent plant life began to draw down carbon from the atmosphere. Photosynthesis turns carbon dioxide into wood and other fibres, and these in turn build up in forest soils. The result of such carbon capture and storage is that the levels of the two main naturally occurring carbon-based greenhouse gases, CO_2 and methane, steadily declined. But around 7,000 years ago an unexpected anomaly kicked in. CO_2 levels went into reverse and started rising.[20] Then methane levels followed suit around 5,000 years ago.

Produced by decomposing vegetation, methane is the major component of swamp gas. Formed in wetlands, it bubbles up when one walks across a bog. Once released into the atmosphere it has a half-life of only seven years before breaking down into CO_2 and water. That is to say, half of what is left of it at any given time will degrade within the next seven years. However, such is its potency as a greenhouse gas that, while it's still around, its 'global warming potential',[21] as measured over a twenty-year time horizon is 84–87 times that of CO_2. Over the more usually considered timeframe of a hundred years, this drops to 28–36 times.[22] Research suggests the possibility that the CO_2 rise may have been caused by Neolithic hunter-gatherers beginning to fell and burn the forests of Europe and Asia. Subsequent herding of cattle, sheep and goats may have stopped the forest regrowth. The methane burst is even more interesting. One school of thought suggests it to have been caused by the rapid spread of irrigated paddy-field wetlands for growing rice across China and South East Asia at that time. This, but probably not this alone, may account for the shift to wetter conditions on the Atlantic north-west of Scotland. It is salutary to think that the loss of the forest at Loch Leurbost and elsewhere might have had some link to agricultural innovation not so very far away from where the Indonesian Papuans originated.

There is, however, a silver lining to this theory, which is called the 'early anthropogenic hypothesis' and was first set out by the University of Virginia climatologist William Ruddiman.[23] Ice age cycles of around 100,000 years are the planetary norm in recent geological history and are caused by rhythmic changes in the earth's orbit relative to the sun. The warm interglacial periods that bring each cycle to a close are brief exceptions to the frozen norm, and usually last for just over 10,000 years. The rise in greenhouse gases 5,000–7,000 years ago may have played some part, together with complex interactions in the earth's orbital cycles, in averting the next big chill.[24] That doesn't sanction colossal fossil fuel emissions, but it does invite the thought that although Loch Leurbost lost its pines, at least it hasn't gone back to semi-frozen tundra.

Sea Level Rise and Storm Effects

It's one thing for the pines to have died and for peat to have grown up around their stumps, but why their disappearance, now slipping beneath the high-water mark, and the loss of the *feannagan*? Global sea levels are rising, caused mainly by the expansion of a warmer ocean in addition to enhanced meltwater runoff from glaciers and polar icecaps. The rate of rise is escalating. From 1901 to 1990 the average was 1.4 mm per year.[25] From 1993, when accurate satellite measurements began, to 2019, the average rate of increase has risen to 3.3 mm and, most recently, to 3.6 mm.[26] In my lifetime, since the mid twentieth century, average world sea levels have risen by about 150 mm, or 6 inches. We see here in global *absolute* mean sea levels one reason that explains the changes visible at Leurbost.[27] However, there are at least three other complex factors to take into account when reading such a shoreline, all of which shed light upon the intricacies of unravelling climate science.

First, storm surges at 'spring' tides, these being the fortnightly very high tides at the full moon and dark moons. If, to cool it down, you blow too hard across a saucerful of hot tea, it spills over the far edge. Similarly, when waves reach landfall in a storm, water piles up and raises the tide above normal predictions. It is striking that tide tables for coastal ports are called tidal 'predictions'. Oceanographers know that weather-related complicating factors can get in the way including, if the wind is blowing offshore, negative surges yielding tides lower than predicted. Local flooding from rivers and runoff can add to storm surges, also known as a skew surge because the reality skews the predictions. In addition, extreme low pressure associated with Atlantic storms reduces the weight of atmosphere pressing down on the sea. This causes a slight upwards bulging of the sea's surface beneath the passing centre of the cyclone. As it faces south-east, towards the mainland, surges measured at Stornoway, just a few miles north of Loch Leurbost, are usually less than a metre.[28] In contrast, on the Atlantic coast of the island, exposed (as in January 2005)[29] to westerly gales of up to hurricane Force 12, and with gusts well in excess of 100 mph, the effect may be considerably higher.

Second, absolute sea level rise and storm surges are compounded by more powerful waves. Over the sixty-nine year period 1949–2017 the average height of winter waves in Scottish and Irish waters has increased by 10 mm a year, more than two-thirds of a metre in total.[30] Big waves mostly wouldn't affect the sheltered end of Loch Leurbost. However, such measurements do serve as a broad indicator of the increasing agitation of the sea. The combined effect is that incoming waves break closer in and penetrate further inland, exhausting their force on land not previously eroded. A sense of how real rising-wave effects are can be seen in how they worry the island's ferry operator. They threaten lifeline services.[31] Typical of anecdotal evidence would be a comment made by Captain Alex

Morrison when he retired at the end of 2019, after forty-five years of skippering CalMac ferries of the Hebrides: 'The weather is more unpredictable nowadays than it was in my younger days. There was more of a set pattern to climate at that time. If you had a depression coming in from the Atlantic it would pass through and the wind veered and eased off, but now it seems to have more of a sting in the tail . . . more vicious than there was in the past.'[32]

Third, *absolute* sea level rise around the world is one thing, but this can be modified locally to a *relative* sea level rise, the latter being adjusted for movements in the earth's crust. It used to be said that although world sea levels are rising, Scotland was relatively safe because the land is continuing to rise after being pressed down into the earth's mantle by the ice age. This rising effect is known as 'crustal rebound' or 'isostatic equilibrium', and with many coastal settlements built on raised beaches, is considered to be a good thing. However, it turns out that not all parts of Scotland are uniformly blessed. To simplify, consider a bowl of jelly. If a weight is placed in the middle, it will sink down but rise up at the edges. Remove the weight, and equilibrium is restored as the middle comes back up and the edges sink back down. The same is true of some areas on the edges of where the ice was thickest.

At Stornoway, measurements suggest that the mean annual rise in sea level is not 3.3 mm, as would be expected from the absolute sea level rise caused by warming across the world, but a disturbingly high 5.7 mm. This appears so, at least as measured over 1992 to 2007 being the period over which the study to which I am referring drew its data.[33] While caution must be exercised in reading too much into observations from such a short time period, it does suggest that the edges of the 'jelly' are settling back down. Isostatic equilibrium in a downwards direction may, perhaps with other local factors, be adding up to 2.4 mm per annum to Stornoway's relative sea rise – the global absolute rate adjusted to local conditions – thus accounting for what is probably by now a full 6 mm per annum of relative sea level rise, this being about an inch in every four years. Together with increased storminess leading to occasional slight flooding in low-lying parts of the town, this cause for concern increasingly finds expression in the local press.[34]

The Three Great Floods

At this point, it will be clear that the science of interpreting a land-scape under change involves multiple factors. Expert help is needed to unravel them. One is always wrestling with inadequacies of data and baselines, but the bottom line is clear enough. All around the island and elsewhere in the world, land is being lost to rising sea levels. A factor like tectonic shift is outside of human control, but anthropogenic climate change is firmly inside our domain of responsibility.

In passing, I must ask my readers from elsewhere to forgive my frequent use of examples from my own bioregion. I speak from what I know best, but it will be plain that many of these points have wider parallels. We shall return to the importance of the Papuans' visit in Chapter 8. Suffice for now to say that they too are worried by the same issues. Many of their villages perch on narrow shelves of land down by the shore. Mountains rise sharply up behind them. If the sea level keeps advancing, they'll run short of ground to cultivate. At Leurbost, we were able to explain what they saw as being a combination of multiple factors. The ongoing restoration of equilibrium following the ice age may be one consid-eration. In Papua, with its abundant volcanic activity, shifts in the relative sea level may have different causes in the earth's crust. It is striking that Indonesia has become the first country in the world to have announced plans to shift its capital city because of relative sea level rise. Jakarta, on the island of Java, will move to Kalimantan, the Indonesian part of Borneo, because the old trading port is suffering from a rise in water levels by up to 28 cm a year in some neighbourhoods. While this is mainly caused by local subsidence in a city of 10 million that has sprawled across former swamps and relentlessly extracts fresh water from below, it is compounded by the rise in absolute sea levels globally.[35]

A few years ago, I was at an award ceremony for young crofters – small-scale farmers and fishers – in the Scottish Parliament. The winner was a fourteen-year-old girl from the Hebridean island of South Uist. She said that twice, in her few years of memory alone, her father had been forced to shift the fence where their land goes down to touch a sandy shoreline. On each occasion he'd moved it several metres further inland. Specialist mapping projections used by planners suggest that much of the arable land down the west coasts of Benbecula and South Uist will be below peak annual flood tide levels as soon as 2030. The sole protection in some parts is just a thin line of sand dunes that, when blown out in storms, must quickly be replanted with deep-rooting marram grass. The worst affected areas include the Isle of Berneray, population 138, where a young Prince Charles once used to go on holiday to plant potatoes incognito. By 2030, nearly half of its 1,000 hectares will likewise be below the flood tide peaks.[36] On the day that a friend in Stornoway emailed me a link to the map that shows all this, I sent it to a friend in Edinburgh, who sent it to a friend in London, who sent it to his friend Rory in Benbecula, who came back and said they 'had a storm surge two weeks ago that came a long way in'. It's happening before our eyes.

There is a legend of the Hebrides, a prophecy, that tells about the three great floods. The first was the primordial inundation of the Book of Genesis, before God had separated the dry land from the oceans, 'and the earth was without form, and void; and darkness was upon the face of the deep'.

The second was, of course, the flood of Noah, brought on because as Genesis informs us, 'the wickedness of man was great in the earth . . . and the Earth was filled with violence'.[37] The truth of this account is proven, or so inferred the storytellers of old, by the presence of animals like hares and deer upon the Hebrides. As the ark went about its long perambulations, it grounded on the peak of Uisgneabhal Mòr in North Harris. Some of the animals two-by-two got off before Noah got the boat refloated on the high tide and carried on to a less important summit, called Mount Ararat.[38]

Then, there will be the third great flood, and that, they say, has yet to come. In this, as an old woman in South Uist told a folklorist in 1869, there will be an 'overflowing of the Atlantic and the submerging of certain places', until a time will come to pass when 'the walls of the churches shall be the fishing rocks of the people' and amongst the resting places of the dead 'the pale-faced mermaid, the marled seal and the brown otter shall race and run and leap and gambol – like the children of men at play'.[39]

In this flood too, another version tells us, the oceans will sweep over, 'and drown all the islands . . . but Iona will rise on the waters and float there like a crown; and the dead who are buried in her will arise dry and so be easily recognised at the Last Day.'[40]

The story is so ridiculous, but its metaphors so redolent and its beauty so profound when understood beneath the surface, that we will leave it there until a brief reprise much later. For now, over the next three chapters, let us address the fundamental science of climate change.

IMPACTS ON THE WORLD OF ICE AND OCEANS

What counts as science? In this and the next two chapters I want to lead into the most recent comprehensive scientific reports. But first, that crucial question, and its context of how we know what we think we know about climate change.

In general, science consists of making observations about the world, setting up hypotheses to explain them and then trying to disprove that same hypothesis by gathering further evidence that might change the picture. If a hypothesis (or theory) proves resistant to being knocked down, the science is said to be 'settled' or robust. If it gets knocked down, because the evidence base no longer backs it up, then it ain't no longer science. If that's persisted with, it's pseudoscience.

Mostly, we only know what we think we know about climate science because of the climate science. Just as we wouldn't go to a neurosurgeon to get the car fixed, or to a mechanic for brain surgery, so climate science depends on reputably published work. This can't be learned by sitting in an armchair with Wikipedia. When somebody challenged the atmospheric physicist Michael Mann on Twitter, making out that climate science was just an easy way to earn a buck, he replied: 'It's really simple. Just double major in Math & Physics, get a masters in Physics and a Ph.D. in Geology & Geophysics, post-doc for a few years, get a faculty position, get tenure, publish a couple hundred articles & a few books . . . Easy cash!'[1]

Science or Pseudoscience?

Precisely because climate change poses deep challenges to all our lives, a powerful industry has been spawned of 'sceptics' or 'deniers'. Lobbyists for the fossil fuel industry use methods reminiscent of the tobacco industry's denial of the link with cancer.[2] Amongst statesmen, Donald Trump has tweeted: 'The concept of global warming was created by and for the Chinese in order to make US manufacturing non-competitive.'[3] As he eyed up political office, he'd have known that jobs win votes more easily than austerity.

And yet, to get our heads around the science is challenging, even with the benefit of a well-rounded education. Separating good science from weak science and denier or alarmist pseudoscience is not easy. If we are not experts, how might we tell what merits trust, or doubt, if looking at a scientific report? Some indicators might help. We can look at an author's or team's qualifications, past track record, funding sources and the institutions to which they are affiliated. We can look at impact factors, these being academic measures of how often a paper, journal or an author is cited by recognised authorities in a field. We can look for signs of poorly supported reasoning or referencing that would imply weak peer review. We can look at the reputation of an author's publisher to try and filter out vanity journals – these have all the airs and graces of real scientific publications, including fancy websites and claims of peer review, but authors pay a fee to publish what has probably been rejected elsewhere. We might be wary of lone wolves if hard-hitting scientific claims are being stridently asserted. And wary likewise of those the trade calls 'silverbacks' – older and once eminent men (as they usually are), who still pronounce with a head-of-department authority on matters over which they're either out of touch, or aren't within their field. Some failings are, alas, not necessarily malevolent, but all too human.

More widely, the complexity of climate science requires us not to cherry-pick a few reference papers, or the latest shock finding to hit the press headlines, but to assimilate thousands of studies across multiple disciplines. That's why there is no substitute for the mutually peer reviewed and consensus work of professional associations. These synthesise bodies of knowledge and present summaries in forms digestible for balanced public understanding. We might think that we know better. We might plough a dogged furrow shouting, 'Galileo proved them wrong!' But if we're not a Galileo, then the fallacy named after him draws us into pseudoscience; and that draws any who might follow us into unreality.

The IPCC and Contested Science

All this is why, for most policymakers, the gold standard of settled science in climatology are the reports of the IPCC, the Intergovernmental Panel on Climate Change. Set up in 1988 by the United Nations and the World Meteorological Organization, its remit is to advise governments on 'the scientific basis of climate change, its impacts and future risks, and options for adaptation and mitigation'.[4] *Mitigation* means action that can be taken to reduce or prevent greenhouse gas emissions. *Adaptation* means action that can be taken to live with the consequences of mitigation's failure.

The IPCC does not carry out its own research. Instead, it collates the published work of hundreds of teams of leading scientists of governments, universities and other institutions across the world. It issues special reports and, every seven years or so, a major assessment report that synthesises the consensus findings of its specialist working groups. In 2008, when Birlinn published *Hell and High Water*, my earlier and very different book on climate change, I based the science on the Fourth Assessment Report of the previous year. By the time of the fifth report in 2014 – often cited as AR5 – the science had massively refined, but most of the conclusions remained unchanged.[5]

I am very well aware that the IPCC's science, like all science, will always be out of date, and that my reliance on its work will be criticised on that account. IPCC cut-off dates for literature submission are usually at least a year before a report is produced to allow time for assimilation into the process.[6] Furthermore, it probably took one to three years for completed scientific work to have been through peer review and have been published in the first place. What can be presented as relatively 'settled science' will therefore be at the very least two years old, and probably more like three to five years old. In the meantime, new work will all the time – especially if it is dramatic – have been splashed across social and mainstream media. However, my personal view is that unless we are recognised specialists in the field concerned we should be very careful how we use new publications.

In the past year at the time of writing I have seen top-level publications suggesting, on the one hand, that the Antarctic ice sheet is more likely to collapse than had previously been thought, and on the other, that collapse is less likely. That's probably not because one or the other team of researchers is wrong. More likely, conflicting conclusions come from looking at things with different methodologies producing different data around processes that remain only partly understood. How to decide between them? By the volume of media noise? Or is the only honest layperson's approach to bite our lips, and wait for the next balanced expert panel appraisal to come round? Mostly, but not always, I choose the latter.

With three decades of history behind it, how well have IPCC forecasts mapped on to later outcomes? Since 1990, with the first of five assessment reports to date, studies of the early scenarios for surface temperatures on earth have shown close adherence to the models. Most appear to work. One review, published in *Geophysical Research Letters* in 2019, compared a range of models of global surface temperature changes that had been produced between 1970 and 2007. (There was no point taking them from a later date, they'd not have had long enough to run). It found that these had been adept at predicting changes, 'with most models examined showing warming consistent with observations'.[7] But forecasting temperature is the easy bit. The impacts get more complicated around such radical uncertainties as Antarctic ice shelf behaviour, the modelling of clouds and the prediction of extreme weather events. The latter, by definition, are historically rare. It's hard to separate the signal from the noise, and therefore, to model and to forecast confidently.

The most thorough review of which I'm aware was the Copenhagen Diagnosis in 2009. Written in 2007, just after the Fourth Assessment Report came out, it found 'that several important aspects of climate change are already occurring at the high end, or even beyond, the expectations of just a few years ago'.[8] Specifically, the melting of polar ice, which early IPCC reports had downplayed in the absence of a robust evidence base, was now proceeding apace. Set against that, though of a much lesser order of magnitude, the same assessment report predicted the demise of Himalayan glaciers by 2035 'if the earth keeps warming at the current rate'. This was not based on adequately peer-reviewed science, is considered to be an overstatement and climate change deniers have relished it ever since.[9] Error apart, the catch-22 for the IPCC is that if researchers over-predict from a weak evidence base, and the climate subsequently under-performs, so to speak, the overshoot will be seized on by politically influential deniers to hammer their credibility. The Fifth Assessment Report in 2014 subsequently corrected for the fourth's under-prediction. However, this usually goes uncredited by those on the alarmist end of the spectrum, who claim that the IPCC plays down the 'real' degree of risk.

Note my recognition that the spectrum has an alarmist end, albeit very much more thinly tapered. Some say it's better not to use such language, that it exposes to deniers, and the merely uninformed or confused, a flank of vulnerability. I disagree. If we are of the inclination that climate change is real and mostly anthropogenic, our task is not to hang out in a bubble with our friends. Our task is to reach out to the deniers and the honest doubters. Which way does that most effectively? Shouting in their faces about how wrong they are? Or opening conversations that can air where each view is coming from? By way of illustration, I have a peculiar sideline. For more than twenty years I've guest lectured to military officers in advanced training at the UK Defence Academy and other European schools of war. I'm asked to put the case for nonviolence.

I don't get traction just by telling them case studies of what has worked. What gives legitimacy is acknowledging the fallibility of us all. I put it to them that we're all at different posts dug in along a long front, that none of us have got a God's-eye-view and that pacifism is not about being passive, it's about sustained courage. 'Never show fear, do show respect,' as a Belfast shipyard worker once advised me about how to handle fearsome situations when unarmed. I tell case studies that show failures of nonviolence – for example, what appears to be the sorry role of Aung San Suu Kyi in Myanmar, as Burma is now called. They're then open to the successes, like the Good Friday Agreement in Northern Ireland or the Jasmine Revolution in Tunisia. And it's a funny thing, but it can end up feeling like they want to help nonviolence to be effective. A few think it's bonkers and will say so. Even then, there can be a bonhomie. But many have a kind of admiration and say, if somewhat ruefully, 'We wish that it could be.' They've seen war at close quarters and have come to understand the spiral of violence.

It's the same with making friends across the fence of climate change. We need to ask ourselves: how best to talk to a Trump supporter? This at the deepest level is not about deniers, though we have to understand them. This is not about alarmists, though we must face that dimension too. This is about the science, and what realities it exposes that can be drawn forth from our shared humanity, irrespective of where we start from on our journeys of understanding.

The UN's Conference of the Parties (COP)

Back to the more prosaic, and the Sixth Assessment Report from the IPCC is due to be finalised in 2022. To feed into its cycle, three 'special reports' or SRs were released in 2018 and 2019.[10] These lay the groundwork for when the leaders of world governments hope to convene in 2021 for COP 26 – the 26th Conference of the Parties. The parties in question are the almost 200 nations that have signed up to the UN's Framework Convention on Climate Change. This, in turn, arose from the UN's Earth Summit in Rio in 1992, and aims at 'preventing "dangerous" human interference with the climate system'.[11]

Glasgow itself is a very interesting place to hold a COP. A study that came out in 2018 found that it contained all of the UK's ten most deprived areas of the past forty years.[12] One of these is Govan, from where I write, immediately adjacent to the conference centre on the River Clyde from where the gathering, postponed from 2020 by the coronavirus, will hopefully take place.[13] Vérène and I moved here because of our involvement with the GalGael Trust, a community group that builds wooden boats, trains in craft skills and holds a hospitable space where people can connect with nature, themselves and one another – the unity of soil, soul and society.

In a revealing cameo of the odd relationships between climate change and poverty, some ten years ago the GalGael and local partner organisations applied for funding from the Scottish Government's innovative Climate Challenge Fund. We proposed a programme that included measures such as community meals, which would reduce individual energy demand by pooling our resources for cooking, heating and conviviality. However, the application form required a calculation of the project's CO_2 emission savings. Whereas affluent areas have some fat to trim, we soon found that cutting much away from nothing doesn't give you very much. But Scotland thrives by imagination. We got the grant in part by cooking up a cheeky case. Our project would indeed cut carbon – by showing richer neighbourhoods some ways of living more with less. Might such be a pattern and example that Glasgow offers to the COP?

The IPCC's Special Reports on Climate Change

Because the IPCC's three recent special reports serve as stepping stones to both COP 26 and, beyond that, the Sixth Assessment Report process, I shall adopt them as the scientific springboard for this book. They're benchmarks of the science as it stood at the end of the second decade of the third millennium. I shall present them here, not in the order of their publication, but in an order that allows me to build up from a foundation in the physical and biological sciences, and then move on to the social implications of trying to constrain global warming to within 1.5°C of pre-industrial temperatures. The reports, with their standard acronyms, are:[14]

Special Report – *The Ocean and Cryosphere in a Changing Climate* – SROCC
Special Report – *Climate Change and Land* – SRCCL
Special Report – *Global Warming of 1.5°C* – SR1.5

Each synthesises the science from across relevant disciplines. Each is built up by international teams of dozens of authors, with hundreds of advisors, drawing on thousands of mostly recently published papers in peer-reviewed journals, and incorporating tens of thousands of pre-publication expert comments. Arguments for and against particular positions are explored and conclusions are expressed in a calibrated language that describes levels of confidence, agreement and likelihood. These confidence levels are always written in italics to stand apart from non-technical usages of such terminology. For example, where an effect is said to be *very likely*, it means that confidence amongst the relevant experts exceeds 90 per cent. *Very unlikely*, means less than 10 per cent confidence.[15] One might be reminded of the former US defense secretary, Donald Rumsfeld, who spoke in one of his better moments of the 'known knowns', 'known unknowns' and 'unknown unknowns – the ones we don't know we don't know'.[16]

The scientific consensus may not be sure of something that is being said, but the reports' honest appraisal of how sure or unsure is a vital aid to clarifying reality. In the summaries that follow, I shall feed in such italicised terms as much as I can without overburdening the text. Also, to avoid making this book read like the academic treatise that it is not, I shall avoid cluttering the text with numerous chapter and verse references. It's easy to find items by searching the reports online.[17] Furthermore, while the science will be important to give a set of bearings for some readers, it might be less so for others. I invite you to feel at liberty to skim sections that may be more dense than meets your pleasure and your need. We can't all be good at everything, yours faithfully included, says his wife. My aim in these early chapters is to build up a secure platform from which to appraise the 'where we're at'. From there, we can better consider the 'where we're going' and in the later chapters, with faltering and trepidation imposed by the situation, the 'how to get there'.

Special Report SROCC: *The Ocean and Cryosphere*

Cryosphere comes from the Greek, meaning the cold or frozen regions of the earth. The report on *The Ocean and Cryosphere in a*

Changing Climate[18] describes measured changes to both the oceans, and to the polar and glaciated regions that store up fresh water over the land. The oceans cover 71 per cent of the earth's surface, glaciers or ice fields cover 10 per cent, so, between them, this report covers four-fifths of the globe. The oceans moderate the temperature of the atmosphere by functioning as a 'sink' for both heat and the CO_2 that holds it in. Since 1970, they have absorbed more than 90 per cent of the excess heat produced by global warming, and the rate of absorption has *likely* doubled since 1993. However, any environmental sink becomes saturated over time. Even the vast ocean's capacity to keep the earth cool will more and more run into limitations in the future.

To discuss possible futures, the IPCC settled on four emissions scenarios, known as Representative Concentration Pathways or RCPs. Representative, because each represents just one of a range of numerous possible pathways. Concentration, because they are benchmarked to given concentrations of greenhouse gases in the atmosphere. And pathways, because they suggest a trajectory that might be taken to achieve a given outcome. These have been consolidated out of more than 1,200 published baseline and mitigation studies.

In 2018, SR1.5 took what it called the 'current' rate of global CO_2 emissions, those at the end of 2017, as being 42 billion tons a year. But, as we've seen, interpreting these figures can be confusing. In 2019 the International Energy Authority made headlines pointing to 'a historic high' for emissions in 2018, but at a mere 33.1 billion tons.[19] Why so low? Because it was speaking only of its area of remit: namely, 'global energy-related CO_2 emissions' or fossil fuel burning. It wasn't counting emissions from land use, or CO_2 equivalents for other greenhouse gases. At the opposite end of things, the 2019 *Emissions Gap Report* from the UN Environment Programme lands in the 2018 emissions at a stomping 55.3 billion tons.[20] Again, why? Because, as befits the UN programme's wide remit, it's counting in the whole shebang, and as CO_2eq. In science – boring but necessary – it pays to read the small print to see exactly what is being measured.

The key thing to remember – for those who like to remember

this kind of thing – is that it is the accumulation of annual *emissions* remaining in the atmosphere that add up to global *concentrations* at any given point in time. That distinction between emissions and concentrations understood, each RCP is defined by a number that is equivalent to the wattage (or degree) of warming caused by anthropogenic greenhouse gases per square metre of the earth's surface at specified concentrations of CO_2 equivalents in parts per million.

RCP2.6 is a low emissions pathway, 2.6 watts per square metre of heating, where CO_2eq concentrations peak around 490 ppm by 2050 and then decline to about 450 ppm by 2100. This is the eco-future of energy decarbonisation and other CO_2 reduction measures. It would keep global warming to within 2°C of pre-industrial temperatures and close to the 1.5°C Paris Agreement's target that will be discussed in the next chapter. By the end of the century, it anticipates warming in a *likely* range of 0.9 to 2.4°C, with a mean (or 'average') of 1.6°C.

At the opposite end of the range, RCP8.5 is the high emissions scenario, 8.5 watts per square metre of heating, where CO_2eq hits 1,370 ppm by 2100 and keeps on going to kingdom come. As a scenario at the extreme end of the spectrum of likelihood it has often been misleadingly described as 'business as usual', but that suggests some coordination between climate science and government policies. Instead, RCP8.5 is a to-hell-in-a-handcart outlook, as if Donald Trump's or the Brazilian President Jair Bolsonaro's climate change denialism was to become a global norm. It projects escalating emissions and little if anything by way of mitigation, with warming by the end of the century in a *likely* range of 3.2 to 5.4°C, and a mean of 4.3°C.

Here we needn't bother spelling out RCP4.5 and RCP6.0, the two in-between. Suffice to say that SROCC uses mainly the lowest and the highest in its discussions to give contrast, though most experts think that the mid-range values are more probable, with warming by 2100 of around 3°C. But note in passing that top-of-the-range 5.4°C figure for the all-stops-pulled-out, devil-on-fire RCP8.5. It's almost completely off the wall, the upper limit of a deliberately extreme scenario, but it still made it into the ballpark.

Melting Ice and Feedbacks

Warming the earth places frozen tundra, glaciers and the ice caps under stress as a range of factors, known as 'polar amplification', combine to exacerbate warming in the Arctic and the Antarctic. Put simply, various feedbacks amplify ice loss, chief of which is that light-coloured surfaces reflect part of the sun's rays back out to space, but darker surfaces are heat absorbing. A classic demonstration of this 'albedo' (or reflective whiteness) effect is a practical science experiment where the sun's rays are focused onto a balloon with a magnifying glass. If dark coloured, it quickly pops. But if white, the light mostly bounces off and there's no such catastrophic conclusion. Sparkling snow or ice, uncontaminated by soot or other particles in the air, therefore keeps the earth cool; but slush or dark blue water absorb heat, which in turn amplifies or 'forces' warming.

As a consequence of such forcing factors, SROCC states that Arctic surface air temperatures *likely* more than doubled the global rate of increase over the most recent two decades for which it has data. Between 2007 and 2016, losses from the Antarctic ice sheet have been treble those of 1997 to 2006. Since 1967, Arctic snow cover has declined by 5.4 per cent per decade.

The projected loss of ice mass from the world's glaciers between 2015 and 2100 is *likely* to be in a range of 18–36 per cent. This matters, because glaciers store water in the winter and release it during summer, supplying rivers and sustaining aquatic life, irrigation and hydropower. It matters culturally too. One of the saddest tokens of 2019 was when villagers in both Iceland and Switzerland held 'funerals' for their glaciers that had died. SROCC has *high confidence* that a range of warming impacts in the Arctic – including abrupt thaw, shifts in the hydrology and wildfires – will impact not just on animals like the reindeer and the salmon, but also 'harming the livelihoods and cultural identity of Arctic residents including Indigenous populations'.

Arctic Permafrost and Uncertainty

Another area of feedback and a cause of deep anxiety amongst

many environmental advocates is the thawing of the Arctic permafrost. Permafrost, which is to say, ground that never thaws out, not
even in the summertime, comprises soils or sediments in lakes and
oceans that was frozen during bygone glaciations, with some being
added over subsequent winters. The Arctic's permafrost alone contains nearly twice the amount of carbon that is in the atmosphere.
Unlike the Antarctic, which is surrounded by open ocean, the
Arctic Ocean is mostly enclosed. It covers 14 million square kilo-
metres – 5.4 million square miles – which is big enough to swallow
Europe one-and-a-half times over. The surface freezes readily in
the winter, not just because of its latitude but also because the vast
northern watersheds of Russia and North America pour into it.
This influx of fresh water renders it the least salty of the world's five
oceans. Some oceanographers regard it more as a sea, an estuary
that feeds into the North Atlantic.

In consequence, for aeons mighty rivers such as the Lena, the
Yenisei, the Ob, the Mackenzie and the Yukon have fanned deep
alluvial deltas over its extensive and shallow continental shelves.
Fallen trees, autumn leaves and pine needles, eroded peat and
whatever else, all wash down and eventually settle in the ocean
sludge. Some of it will join the permafrost and be stabilised, but
much will slowly bubble out as CO_2 and methane as microbes set
about the slow process of anaerobic digestion – also known as 'rot'
– in a reversal of the chemistry that photosynthesis first brought
about.

This is all a normal part of nature's process. The question posed
by climate change, is the extent to which it might be escalating
abnormally. SROCC's jury remains out. It finds only '*medium
evidence* with *low agreement* whether northern permafrost regions
are currently releasing additional net methane and CO_2 due
to thaw'. A briefing that the Royal Society issued at the end of
2017 to summarise the state of climate science sounded a more
dismissive tone towards the methane threat. It said: 'There is little
evidence of a significant increase in emissions from the Arctic,'
adding that the geographical distribution of atmosphere methane,
together with source analysis from its carbon isotopes, points the
finger more to tropical sources, as well as to fossil fuel use and

methods of extraction.[21] Increasingly, the finger points to shale oil and fracking.[22]

With *very high confidence*, SROCC anticipates widespread permafrost thaw this century. Whereas there may not be cause for alarm over emissions yet, that doesn't discount concerns about the future. On high emissions scenarios there is *medium confidence* of a 'cumulative release of tens to hundreds of billions of tons of permafrost carbon as CO_2 and methane to the atmosphere by 2100'. This will have 'the potential to exacerbate climate change'. Why such bland language for what sounds like a no-brainer? This is where point must often be offset by counterpoint. Additional plant growth will partly mitigate the emissions increase, for three main reasons. The boreal forests and other habitats will extend northwards, the growing season will be warmer and therefore longer, and CO_2 is a fertiliser. This growth will fix carbon down into ground vegetation cover and the soil, but with *medium confidence*, these counterpoints will not be sufficient to compensate for all the carbon lost elsewhere in the polar ecosystem. We can glimpse here just how stilted one-sided arguments about climate change can be from whichever side of the debate they're made. There is no substitute for balance. That said, the balance says that only by cutting greenhouse gas emissions, and thereby stabilising and preferably heavily reducing atmospheric greenhouse gas concentrations, can very serious future risks be averted.

A major problem with assessing and modelling in the Arctic is that too few baseline studies exist to be sure of what is being compared with what. SROCC describes deep divergences in both the understanding of key variables and in the conceptual frameworks used in their modelling. For example, the northerly migration of boreal scrub and forests will draw down carbon, but it is hard to predict how much of that benefit will be offset by dark green ground cover in place of snow, reducing the albedo and thereby exacerbating warming. It is because there are so many such uncertainties that SROCC takes Arctic permafrost emissions as one of three case studies of 'deep uncertainty'. The other two are the Antarctic ice sheet, and what it calls 'compound risks and cascading impacts', these being what happens when the dominos fall down. SROCC is very aware of how unsatisfactory it is to not have answers to everything. It states: 'With stakeholder needs in mind, scientists have been actively engaged in narrowing this uncertainty by using multiple lines of evidence, expert judgment, and joint evaluation of observations and models.' The uncertainty is starting to reduce, but only 'across some but not all components of this problem'. In short, we may not have the clarity that we might like, but at least there's honesty about the darkness.

An Arctic Methane Time-Bomb?

Many readers of a book like this will be well aware that Arctic albedo and methane have become headline triggers of climate change concern. In Chapter 6 we will discuss the fears held by some campaigners around the possibility of imminent social collapse, mass dieback of our species and even near-term human extinction. These anxieties primarily bounce off the worry that a massive methane 'time-bomb' could set off an unstoppable chain reaction of warming. The more methane released, the hotter the world would get. That in turn would cause still more to be released, like a nuclear reactor gone critical with meltdown to the core.

In 2012, Professor Peter Wadhams – a reputable polar physicist, but one who has proven fallible in forecasting Arctic sea ice loss – summarised his worry in these words:

> Already we are seeing consequences from these changes.
> The new large area of [Arctic] open water warms up to
> 4–5°C during summer, which not only delays the onset
> of autumn freezing but also warms the seabed over the
> shelf areas, helping to melt offshore permafrost. One con-
> sequence of this melt is the release and decomposition
> of trapped methane hydrates, causing methane plumes
> which have global warming potential. Already such
> plumes have been directly observed in the East Siberian
> Sea . . . [T]here have been warnings that a major methane
> outbreak may be imminent, with release from offshore

permafrost melt being joined by releases from the active layer under the tundra . . .'[23]

Wadhams was referring to the work of a mainly Russian team led by Natalia Shakhova and Igor Semiletov, experts in the East Siberian Arctic Shelf.[24] SROCC cites one of their papers, but alongside others that in aggregate bring out the challenges of separating signal from noise. One of these suggests that actual methane releases may be six to ten times smaller than Shakhova's team had estimated.[25] All that SROCC can honestly conclude is that at the moment, there is *low agreement* amongst the experts around such matters, and that Shakhova's paper has 'widened rather than narrowed the uncertainty range' as to what may or may not be going on.

To most climate scientists, Arctic methane – including in its hydrate (or clathrate) that forms as a kind of ice on the seabed and in permafrost – is not the most pressing near term concern. This reflects in the very few words that SROCC affords the matter. Nevertheless, methane alarms alarmist bloggers.[26] We have already seen that its global warming potential is tens of times that of CO_2, depending on the time period over which its seven-year half-life in the atmosphere is taken into account. However, this acute effect is not persistent. The persistent concern is the CO_2 into which the methane oxidises. This makes most experts dismiss the 'time-bomb' hypothesis on any of the likely scenarios for this century. In a post called 'Much ado about methane' on RealClimate, the climate scientists' discussion site, David Archer examined the tendency of methane 'to really get people worked up, compared to other equally frightening pieces of the climate story'. He summed up the situation, saying: 'When methane is released chronically, over decades, the concentration in the atmosphere will rise to a new equilibrium value. It won't keep rising indefinitely, like CO_2 would, because methane degrades while CO_2 essentially just accumulates . . . Actually, releasing CO_2 is a point of no return if anything is . . . Conclusion: It's the CO_2, friend.'[27]

Similarly, in a recent 'fact check' review on climatetipping-points.info, another scientist-run site and one that was set up with seed funding that came via the UK's Engineering and Physical

Sciences Research Council, David Armstrong McKay, who models nonlinear biosphere-climate feedbacks – in plain language, 'tipping points' – at the Stockholm Resilience Centre, lucidly explains the myths. Even if the methane did 'go off' abruptly, the temperature rise would be relatively minimal and transitory. This leads him to conclude: 'The most likely situation then is one of a gradually growing chronic leakage of additional methane and CO_2 from the Arctic over the coming decades and centuries, rather than an abrupt "methane bomb". This will act as a gradual amplifier of human-driven global warming, making staying within the Paris Targets of 1.5–2°C even more challenging and urgent.'[28]

As both Archer and McKay make clear, these are not arguments for playing methane down. Rather, they're counsel not to play it up by shortening the timescales in which the risks should be situated given the relatively short half-life of the gas. If we exceed the consensus science, there is a danger that we'll jump to what might be the wrong solutions. Indeed, one body, the Arctic Methane Emergency Group, has already directed its concerns towards the British government in an attempt to leverage the case for geoengineering.[29]

Slowing the Atlantic's AMOC Current

The Atlantic Meridional Overturning Circulation (AMOC) – of which the commonly known Gulf Stream forms a component – is a system of ocean currents that move warm water at the surface from the tropics northwards, and moves cold water at depth from the North Atlantic southwards. This keeps British winters relatively mild and prevents ports from freezing over such as they do at the same latitudes in Newfoundland or Hudson Bay.

As warm water carried along in the AMOC heads north, it cools and becomes more dense. Evaporation along the way makes it saltier, and this also adds to the density. On approaching the upper latitudes, it starts to sink and overturn at depth, causing a 'pump' effect. This propels it back southwards along the meridians, the earth's lines of longitude, thus why it's called 'meridional' overturning. Most of it resurfaces around Antarctica from where it joins up with wider patterns of the world's interconnected ocean circulation. However, the influx of fresh water from melting ice, together with the Arctic's waters being warmer due to climate change, reduces the current's density as it reaches higher latitudes. This could reasonably be expected to cut down its tendency to sink. It would thereby slow down the pumping mechanism, and in so doing, reduce or even stop the warmth transported to north-east Atlantic maritime countries. SROCC has *medium confidence* that the AMOC has already weakened since the nineteenth century. Under even the most optimistic emissions scenarios, it is *very likely* to weaken further in the twenty-first century, although collapse is *very unlikely*: at least, not until 2300 on a high emissions scenario. Under those conditions, collapse would become *as likely as not.*

With mostly *medium confidence*, any emissions scenario will cause the AMOC to slow down. Such is the interconnectedness of world weather systems, and the importance of ocean currents for moving energy around, that the list of consequences is very long. On the upside, there would be a reduction in Atlantic tropical cyclones. On the downside, we might anticipate more North European storms and disrupted temperatures, and further afield, changes to rainfall patterns stretching from the Sahel to South Asia. It has also been speculated that increasing atmospheric temperatures might offset the cooling of British ports, but as that is a very regional concern, it is not the level of detail that SROCC mostly addresses.

Sea Level Rises

As the oceans warm, their volume expands and causes the sea level to rise. Meltwater runoff from the cryosphere adds to the process and has recently overtaken expansion to become the greater factor. The rate of rise can be assessed from tide gauges. The records of some European ports go back to the eighteenth century. However, their accuracy is often compromised by local conditions, including movements in the earth's crust and changes to water flows due to silting, dredging, extraction from rivers and harbour construction works. As such, the baseline from which reliable, planet-wide satellite measurements began, is only 1993. Since then, the rate of global sea level rise has averaged 3.3 mm a year, but the rate of escalation is such that between 2006 and 2015, it *very likely* rose by 3.6 mm per year.

SROCC describes this as being 'unprecedented over the last century'. Worldwide, as figured out from averaging a great many tide gauges, the sea has risen by 0.15 m (150 mm, about 6 inches) during the twentieth century. Looking ahead to 2050, there is *medium confidence* of it rising in the range of 0.24–0.32 m, in other words, roughly double the twentieth century's entire rise in just some thirty years. Looking still further ahead to 2100, the *likely range* is 0.43–0.84 m, but on a high emissions scenario such as will continue unless the world effects strong greenhouse gas mitigation measures, the upper *likely* range for 2100 is 1.1 m, and a rise of up to 2 m 'cannot be ruled out'.

One study included within the ballpark of credibility by SROCC pushes the range on an RCP8.5 scenario to a rise of between 1.84 and 2.46 m by 2100. Beyond this century, there are few modelling studies to suggest what might come to pass. Uncertainties such as the Antarctic ice sheet make it challenging to predict too far ahead. SROCC simply leaves it as a matter of *high confidence* that 'sea level will continue to rise for centuries and will remain elevated for thousands of years', with some simulations speculating that it could be 10 m per millennium while the ice lasts if the CO_2 levels are allowed to rise high enough. If all the world's ice were to melt, the global sea level would increase by possibly as much as 80 m, about 260 ft.[30] While useful to show how much fresh water is stored in the cryosphere, that scenario is not within the ballpark of credibility because Antarctica has an average elevation of 2,500 metres.

From what has been discussed in this section, we can see that while there's still some time to adapt to a habitable future, there's no time left to hang around the beach. However, there is good news for low-lying New York. Taxpayers might not have to foot the bill. For sure, the US Army Corps of Engineers are examining options to protect it with a $200 billion sea wall. A pressing argument in favour is that Hurricane Sandy alone set the city back $19 billion. But President Trump, who knows a lot about walls, tweeted that this would be a 'costly, foolish and environmentally unfriendly idea'. He said that it probably wouldn't work anyway, and would

look terrible. He has a cheaper option that should see his generation out. You couldn't fake the empathy. The tweet signed off:
'Sorry, you'll just have to get your mops & buckets ready!'[31]

The political effects can indeed be tragicomic. I don't know
what it is about parts of the USA, but in 2012, concerned about the
values and insurability of their so-called 'real' estate, a consortium
of North Carolina property owners and businesses successfully
pushed House Bill 819 through the state legislature. This forbade
both state and local agencies from using any definition or forecast
of sea level rise, except the one stipulated by their own Coastal
Resources Commission. Conveniently, it also restricted the planning horizon to just thirty years, so you weren't to think beyond
your own generation.[32]

Such bullheadedness would have impressed King Canute of
England in the tenth century: reportedly, he tried by regal command, but failed by will of God, to turn back the tide. Evidently,
he never thought of trying legislation.

Storms and Natural Variability

As fits with what we saw happening in around the Hebrides,
SROCC expresses *medium confidence* that extreme wave heights
have increased in the Atlantic. There is *medium confidence* that
precipitation – mainly rain and snow – has globally increased and
high confidence that tropical cyclones are causing more extreme sea
level events, with adverse cascading consequences. However, there
is only 'emerging' evidence with *low confidence* that category 4 and
5 storms have increased globally, notwithstanding some pronounced
regional observations such as the multiple Atlantic hurricanes in
2017.

The difficulty in being sure of what is happening with both
the frequency and severity of storms is attributed to 'the large
natural variability, which makes trend detection challenging'. As
with many aspects of climate change, one gets the sense that scientists – like seasoned mariners in this case – feel that conditions are
worsening, but lack sufficient historical baseline data to say so with
the confidence that their careful profession demands.

Ocean Acidification

Another burning issue is ocean acidification. The ocean serving as a 'sink' has *very likely* soaked up 20–30 per cent of human CO_2 emissions since the 1980s, heavily protecting the world from the full potential of atmospheric warming. However, not only does a sink start to saturate and become less effective over time, it also changes in its qualities. Sea water, which is naturally alkaline, becomes less alkalescent because CO_2 mixed with water becomes weak carbonic acid. Such acidification is why a glass of soda water tastes slightly tart even after going flat.

The sea will never literally become acid. The term 'acidification' only means a movement in the direction of becoming less alkaline, by edging towards acidity. This is detrimental for calcifying organisms such as corals, shells and calcareous phytoplankton (plant plankton), all of which capture CO_2 and bind it into marine sediments, eventually to become rocks. Acidification also harms the early life stages of other marine organisms. Such challenges attack the base of the marine food chain. SROCC holds that there is *high confidence* that harm is already occurring. It is *virtually certain* to exacerbate towards 2100, with the alkalinity of the ocean *virtually certain* to decrease by 0.3 pH units over the remainder of this century. As pH is measured on a logarithmic scale – that is to say, one that bends very sharply instead of being linear – that sounds like a tiny amount, but it actually represents a very big shift. In everyday language, it doubles the ocean's acidity.[33]

This is a major concern that gives the lie to those geoengineering proposals that would purport to tackle global warming by reducing the intensity of incoming sunlight. Turning down the sun by measures such as mirrors in the sky would be the easy bit. But acidification would continue unabated. It would drive some or many marine species to extinction, and reduce the food supply to what was left. Marine life fixes CO_2, especially shellfish, which are surrounded by an exoskeleton of calcium carbonate. To compromise the wellbeing of mussels, whelks and oysters, is to bring about a stomach upset on a planetary scale.

Marine Heat Waves and Deoxygenation

Adding to the woes of acidification, it is *very likely* that marine heat waves have both doubled in frequency and increased in intensity since 1982. With *high confidence*, such regional rapid excessive heating – hand-in-hand with a deoxygenation of the ocean's surface layers that warming also causes – have had a range of ill effects. SROCC states that these are already impacting marine productivity, human health, tourism and coastal economies.

One *high confidence* example of this, is an increased incidence of harmful algal blooms. There is *very high confidence* that, worldwide, coral reef bleaching is taking place. This degradation of the reefs shows in a loss of colour, a loss not least of beauty to the world, caused when the polyps are under stress and are pushed to the edge of their biological limits.

The bottom line is both *high confidence* and *high agreement* amongst the relevant experts that the 'seven seas' of the world are afflicted by a triple whammy. What we see here is that greenhouse gas emissions not only cause global warming, they also introduce a variety of other detrimental earth system changes. SROCC sums it up: 'The combined effects of warming, ocean deoxygenation and acidification in the 21st century are projected to exacerbate the impacts on the body size, growth, reproduction and mortality of fishes, and consequently increases their risk of population decline.'

Ocean Stratification and Nutrient Flows

Finally, and by no means exhaustively, another problem is 'density stratification'. We saw with the AMOC that water that is either warmer or less salty, weighs less per unit of volume than that which is cold or laden with dissolved salt. Beyond certain temperature thresholds, mixing no longer takes place easily. A warmer layer, or one that is less saline (perhaps because of runoff from the melting icecaps) will float like a thickness of oil on top of what is underneath. Such stratification inhibits the upwelling of nutrients from the seabed to the surface, where light can penetrate and photosynthesis can take place for the flourishing of seaweeds, phytoplankton and other forms of algae. Being the bottom of the food chain, these provide both habitat and nutrition for nearly all marine life higher up, from zooplankton – made up of minute sea creatures – to the great whales.

Unpolluted tropical waters are usually crystal clear because density stratification inhibits the transport of nutrients that nourish phytoplankton. At the opposite extreme, upwelling cold waters bring nutrients to the surface. This happens in such places as the Grand Banks off Newfoundland. Before ruinous industrial fishing caused its catastrophic collapse in the early 1990s, the Banks had been the greatest fishery the world had ever known. The ice-cold Labrador Current comes south, meets the warm (and properly called) Gulf Stream part of the AMOC heading north, creating much fog and introducing nutrients that, as the phytoplankton gets going, can turn the sea a soupy green.

Such terms as density stratification can come across as very abstract. Yet on the water, they can be quite palpable. As a boy I would go fishing with old Finlay Montgomery of Ranish, out on an arm of sea, Loch Grimshader, that neighbours Leurbost. There's a river flowing in, the force of which is softened by a bag end in the bay – a bulge that then narrows before it widens – which therefore leads to little mixing. When the weather has been calm, meaning no turbulence from waves either, the fresh water stratifies across the sea. It freezes readily in cold winters, white-frosted like some offshoot of the Arctic. When you're out in the boat the surface water tastes brackish, sometimes almost good enough to drink.

We'd let down our mussel-baited handlines 6 fathoms, about 12 metres. Until commercial fishing robbed us of it in the early 1970s, we'd pull up saltwater species: haddock, whiting, codling and even dogfish – a small shark with a skin like sandpaper. That was the experience of density stratification in real living. Underneath the layer of brackish freshness, the salty ocean pulsed its rhythm to and fro with every tide.

CLIMATE CHANGE ON LAND AND HUMAN LIFE

The IPCC's special report on climate change and terrestrial ecology – the ecology of land as distinct from the waters – overlaps with the science of the oceans and the cryosphere that has just been summarised. In this chapter, I will therefore soft-pedal on the natural ecology and focus on the human ecology that is so rich in the special report, *Climate Change and Land* (SRCCL).[1] This covers the 20 per cent or so of the earth's surface that is under neither ice nor water. The report's home web page highlights that 40 per cent of the coordinating lead authors were women and 53 per cent from developing countries. It was launched in August 2019 with a summary tweet:[2]

> Land is where we live.
> Land is under growing human pressure.
> Land is part of the solution.
> But land can't do it all.

At present, human beings make use of more than 70 per cent of the earth's land. Its use accounts for 13 per cent of global emissions of CO_2 and 44 per cent of methane, the remaining lion's share of greenhouse gas emissions being from industry and fossil fuels. Land use also accounts for 82 per cent of anthropogenic emissions of nitrous oxide, this being a potent greenhouse gas that accounts for 6 per cent of global warming and comes mainly from nitrate fertilisers.[3] As we will see later, much is made of the potential to draw down CO_2 and combat global warming by reforesting the earth and rebuilding carbon into soils. That capacity shows how much has been degraded. Leaving aside the technological twist of bioenergy with carbon capture and storage, we'd just be putting back what we've already taken out. That might hint at just how much we've upset the balance between the atmosphere and land.

Special Report SRCCL: *Climate Change and Land*

SRCCL has *high confidence* that the average temperature of the air over land surfaces has risen by nearly double the global average of both land and sea combined. Between 1850 and 2015 the global mean surface temperature of land and ocean combined went up by 0.87°C, whereas over land alone the rise was 1.53°C. This is because the sun's rays warm the immediate surface of the land which in turn warms the air above, whereas on the ocean they penetrate the surface, from where more of their energy is absorbed and retained as a heat 'sink'.

To put those figures into some perspective, Met Office statistics for 1961–90 give the average temperature of Scotland as 7°C, and 8.3°C across Britain as a whole.[4] That's averaged over day and night, summer and winter, and not to be confused with tourist websites that might give July means at noon, or January ones at midnight, depending on whether it's for sunbathing or snowboarding. Given that the freezing point of water is 0°C, a difference of a degree and a half over land is the difference that makes a difference between, say, whether it freezes or it thaws. Whereas a reduction in winter misery might seem to be a mercy, frost is the farmer's friend. It kills off bugs, and soil churn – the heaving effect that is caused by the expansion of water as it turns to ice – is nature's ploughshare, letting air down to the roots and releasing nutrients by splitting stones and rocks. Without a 'good hardy frost' more fertilisers and biocides would be used.

With *high confidence*, warming has increased the frequency and intensity of heat waves and, with *medium* confidence, that of droughts in many parts of the world. Since 1961 the annual area of drylands under drought has increased by slightly over 1 per cent a year. Multiple factors will lie behind that, but SRCCL has *high confidence* that current levels of global warming are associated with moderate risks of drought and its knock-on effects. Water stress in turn contributes to poverty, conflict and migration, so why only 'moderate' risks? In part, because one area's loss can be another's gain. Such moderating language, however, belies the disruptive effects of change, in addition to which the threat assessment rises to *medium confidence* of 'high' risks at 1.5°C, and 'very high' at 3°C. Trying to weigh up these issues is fraught with complex interactions. For example, warming can increase drought which increases fire risks which burns forests which puts more CO_2 back into the atmosphere which further increases warming. But multiple human factors can multiply or reduce the impacts. I'll give an example from experience.

Drought, Fire and Flood

In the summer of 2010, Vérène and I were returning to her parents' home at Montaud, outside Montpellier in France. We'd been a few days in Robert Louis Stevenson country in the nearby Cévennes mountains, and as we approached we could smell burning, then saw the *pompiers* out drying their fire hoses, and then could see that many miles of lightly wooded land had been completely blackened. On getting to the house we found that her parents had been woken by the emergency services in the middle of the night. They said to pack up just a single suitcase with vital medicines, documents and valuables, and to flee immediately. Their home escaped by just a few hundred yards. Others further up the lane were either razed or damaged.

The background factor was severe drought. This echoed SRCCL's *medium confidence* that the frequency of droughts and the length of the fire season has already increased in the Mediterranean, and that with around 2°C of warming, there'll be a 50 per cent increase in the area destroyed each year by fire. But there is more to the ecology of fire than climate alone. The direct cause of this particular catastrophe was that a disaffected youth had gone round on his motorbike, sprinkling petrol and dropping matches at several locations. Here we glimpse that planetary health cannot be separated from mental and wider social health.

The counterpoint to drought is flood. Turn up the heat on a saucepan and the kitchen fills with steam. What goes up must come down, and warmer air can hold more water vapour. The same applies to the planetary thermostat, but with variable results. With *high confidence*, both the frequency and intensity of extreme rainfall are expected to go up in many regions, thereby causing flooding. But it's not all bad news. Shifts in rainfall patterns that cause drought and vegetation browning in one region can bring greening to another. Globally, satellite observations give *high confidence* that the world is currently undergoing more greening than browning. We see there one of the counter-intuitive consequences of climate change.

On which matter, Naomi Seibt, the German teenager with an extensive far-right support base and nicknamed the 'anti-Greta Thunberg', copies Donald Trump's claim that global warming was rebranded to climate change because its predictions weren't working out.[5] Not so. I recall nearly fifteen years ago one of my ecological mentors, Tess Darwin, who then worked for the government agency Scottish Natural Heritage, putting me right on this point. She said that their scientists' recommendation of good practice was to speak of climate change rather than global warming. 'Warming' makes people think that the effect will be uniform across the world, and wouldn't that be rather nice, given Scotland's dreich weather and average temperature of just 7°C? However, the effects of warming on the world's climate include many consequences that are counter-intuitive. For example, by both disrupting and increasing turbulence in the global climate system, some regions can be expected to become even cooler. Scotland would be one of these were the 'cold blob' or 'blue blob of death' in the Atlantic below Greenland to expand its present extent – this being the present regional trend of ocean cooling that is probably caused by global warming slowing down the transport of tropical warmth in the currents of the AMOC.[6]

Set against that advice, polling carried out by the climate change denial lobby suggests that to speak of climate change is less frightening to the public than warming.[7] Some climate advocates therefore suggest that we should speak of 'global heating' instead, so that we properly convey the danger. In short, there has been no 'rebranding'. Each of these terms can be appropriate and the choice probably depends on context.

Pests, Pollinators and Disease

It was on another annual visit to her parent's place, at the same time of year as we usually go but some six years later (give or take a year on either side), that Vérène and I were back up in the Cévennes where we witnessed a wonder of nature. There had been a mass hatching out of butterflies. As we'd walk the mountain trails, from every bush a cloud of perhaps a couple of dozen would rise up and of many different sizes and colours. I had never seen such a profusion before, and butterflies are symbols of joy, hope and the soul. Ecologists, however, can have a reputation for never being happy. Part of me was left wondering as to the plight of local farmers' crops, many of whom work to organic or *biologique* standards and avoid the use of pesticides. It did fleetingly cross my mind that where there was a host of butterflies, there must earlier have been plague of caterpillars.

Plagues of whatever type have always blighted the affairs of humankind. The East Africa desert locust crisis that began in 2018 may well have its origin in climate change affecting the breeding conditions, but otherwise it echoes cataclysms that run through the Quran, the Bible and other early literature. How might the shifting circumstances of today impact upon the 'balance of nature' that otherwise tends to keep such excesses in check? SRCCL tells that climate change is altering the interactions between pests and disease, with *robust evidence* that these have already responded to climate change. Both in terms of pest outbreaks and diseases for which insects are the vector – the means of transmission – this is already affecting crops, livestock, people and wild nature. Insects can cause harm both directly in their bites, stings and what they eat, and indirectly as the vectors of sickness. Examples include Lyme disease, which is spread to humans by ticks, and bluetongue virus, to which sheep are especially vulnerable and which is spread by biting midges. SRCCL cites evidence suggesting that the existing ranges of about 49 per cent of insects will be reduced by half by 2100. If the range cannot expand elsewhere in compensation, that means that species will become extinct. The swings and balances of all this mean that, overall, there is *medium confidence* that pest and disease exposure will on average increase with climate change. However, some risks in some regions may decrease. For example, a reduction in rainfall and therefore humidity can reduce the viability of fungal pathogens (as disease-causing organisms are known).

When climate change shifts the ecological balances between pests, predators and pathogens, some species can be favoured over others, and their growth stages in relation to one another can get out of kilter. A change in the timing of pollinating insects can miss the flowering times of plants and cause the failure of a harvest. Fewer frosts in the spring may be good for blossoming fruit trees, but not if the conditions are so mild that pests don't die in winter. Insects, birds and bats all pollinate plants. SRCCL states that at present there is only *limited evidence* that their role in pollination is being impacted, *medium evidence* that there will be impacts in future, and *high agreement* that both pest and disease pressures are likely to change. What can we make of such a seeming contrasts of confidence? My reading is that it's a bit like the controversial 'not proven' verdict in Scots law. As some would colloquially interpret the principle: 'We know you did it, but we can't prove it'.

The IPCC does not directly discuss recent 'insect Armageddon' reports of mass declines.[8] This might seem strange, given its emphasis that pollinators are crucial to global food security, accounting for up to 35 per cent of crop production by volume. Here again, like we saw with Arctic methane, popular perceptions and the consensus scientific evaluation can be at odds. A good example of such mismatch between popular perceptions and measured evidence is a key paper published in *Nature Ecology & Evolution*. It came out too recently to have been included in SRCCL. Here scientists involved with the Rothamsted Insect Survey, the world's longest-running insect population study, monitored the weight (or 'biomass') of moths trapped in a consistent manner over a diversity of habitats over the past fifty years. Moths are a good indicator of the health of ecosystems. They can be easily trapped by being drawn to a bright light at night, their caterpillars rely on the availability of suitable plant species down the food chain, and the weight of moths produced sustains such predators as birds, bats and amphibians up the food chain.

The Rothamsted study notes that nearly all existing estimates of changes in biomass lack a long-term consistent baseline by which to compare the present with the past. With half a century to go on, it finds that the British moth population has fluctuated up and down, and it postulates that short-term weather variations rather than longer-term climate averages may be a leading factor. While they found a worrying gradual decline in moth biomass since 1982, there had been a sharp increase up until that date. This was consistent across both different species and habitats. Most astonishingly, when the compared the first decade of the study (starting in 1967) with the last (ending in 2017), they saw a 2.2-fold net *increase* in the weight of moths captured. Within the specific framings of their study, this means that claims of 'insectageddon' are 'not supported'. They concluded that, while there are no grounds for complacency given the gradual decline since 1982: 'The increasingly widespread view that insect biomass is collapsing finds little support in what is perhaps the best insect population database available anywhere in the world.'[9]

Other evidence that challenges an overly black and white picture of insect apocalypse is the rare positive news that between 1999 and 2017 the UK Bat Index, a measure of the relative abundance of eleven of Britain's seventeen breeding species, increased by 42 per cent. All British bats eat insects, and according to the Bat Conservation Trust, a common pipistrelle can eat more than 3,000 small ones in a single night's flitting. Does this mean that other studies have got it wrong? One German study over a twenty-seven-year time span to 2016 measured a decline of over 75 per cent.[10] However, as the Rothamsted researchers point out, studies conducted after 1982 might pick up on the decline but not on the earlier peak. That said, who is to know whether the peak was not itself an anomaly, like Vérène and I witnessed that late summer in the Cévennes? A 2013 review by Butterfly Conservation painted a mixed and complex picture. It reported worrying declines in moth biodiversity – the number of different species, as distinct from their total weight – but with some of the losses in the south of Britain being offset by gains in the north. This geographical drift, they suggested, is most likely down to climate change.[11]

In reviewing such studies, I have been less able to shed a light of clarity than light on complexity. What might such a confused sea mean for the IPCC? It means simply that the next time a working group sits down to evaluate such evidence, studies such as Rothamsted, the German one, and others involving measuring car windscreen splats will be factored in with many others from around the world. Levels of *agreement*, *likelihood* and *confidence* will be hammered out amongst people who are suitably qualified to balance up, as best they can, both conflicting and convergent evidence. A significant change in an insect population may be down to climate change, or it may be caused by other factors such as pesticides, pollution, habitat loss due to agriculture, predator dynamics or translocated pathogens – that is to say, ones that have been moved from one part of the world to another. As SRCCL sums it all up: 'How complex systems respond is highly context-dependent.'

We might throw up our hands in exasperation at such failure to clearly point fingers of blame. We might exclaim, as I have heard people do at public talks: 'Yes, but it's *everything*!' After all, in ecology everything is interconnected. But overgeneralising doesn't help when seeking to identify specific 'biosecurity' measures that might ameliorate the risks of spreading pests and pathogens, whether with or without climate change. When an airliner comes in to Australia, and before the passengers are allowed to disembark, the crew open all the luggage compartments and pass down the cabin spraying a heavy mist of insecticide. A government website explains that such 'aircraft disinsection' is required by law because, 'Australia is largely free of mosquito vectors that transmit serious diseases including dengue fever, chikungunya, Zika virus and yellow fever.'[12] That's what biosecurity looks like in real life. The consequence of its lapses are made plain if one considers translocated tree pathogens, such as the fungi that cause Dutch elm disease (with a little help from a beetle) and ash dieback.

While I am not an entomologist, I have wondered whether translocated insect pathogens could be a hidden factor in some reported insect declines. A beetle in our hiking gear when we return from France, for example, could bring in a beetle virus to which the local species have no resistance. Those much-vaunted 'bucket lists' could have more than just bad carbon karma lurking in the bottom. It is to the human effects of such globalisation that we now will turn.

Epidemics, Pandemics . . . and COVID-19?

Malaria is transmitted by the female Anopheles mosquito, and SRCCL cites the World Health Organization anticipating that exacerbation of the illness, attributable to climate change, will lead to an additional 60,000 deaths each year by 2030. Unexpectedly, that falls to 32,000 extra deaths by 2050. Why such a rise and fall, notwithstanding the worsening backdrop of global warming? I looked up the WHO report that the IPCC had cited. At first, I thought that the latter must have made a typo. If so, their Geneva office maintains a desk to which possible mistakes can be reported. What's more, I know that it is staffed because while writing this book I thought I'd found an error in one of their reports. In the course of following it up, and before realising that the error had been mine, it was reassuring to find that a real person responded warmly from the other end.

With the malaria, then: had the 32,000 dropped a zero? Should it have been 320,000? Not so. The WHO's explanation is that the illness is related not just to the *Plasmodium* parasite in mosquitoes, but is also a disease of those made vulnerable by poverty. Let me illustrate that point from visceral experience. In 1978, when I was a volunteer living in the sago swamps of Gulf Province in Papua New Guinea, I had malaria several times, even though I took the weekly prophylactic drugs. In that part of the world, it would likely have been of the killer *falciparum* variety. The symptoms would come on within a few hours like a violent dose of flu as the parasite burst open my red blood cells. There'd be one locomotive of a headache, violent vomiting on an empty stomach, I'll not mention the other end, a high fever, exhaustion and gnawing aches down to the marrow of the bones. Thankfully, in those days, a knockout treatment starting with 6 chloroquine tabs would lay it lower than its victim. I was only 22, fitter than I've ever been, and would recover within a few days. However, the following year I didn't have a single recurrence. Why not? Most likely, because previously I'd been living amongst the local people in tight-packed conditions that gave a high infection base, and with inadequate fly wire protection on many of the windows of our rickety accommodation. In the second year, we'd built a new school and moved into well-protected teachers' houses. The contrast was dramatic as my health responded to the move up-market.

By modelling malaria alongside economic forecasts, WHO concluded that, 'climate change has much weaker effects than GDP per capita increase'. The one plays off against the other. That said, there is no cause for complacency over climate change and health. When the WHO's epidemiologists added up the estimated additional mortality attributable to future climate change, including not just malaria but also diarrhoea, malnutrition and heat exposure, they concluded that the world will see 250,000 more deaths a year by 2050 than in 2030.[13]

What about emergent diseases, that can have a terrifying virility – both in terms of morbidity (sickness) and mortality (death) – until herd immunity is built up in the host population? An epidemic is the temporary prevalence of a disease, and a pandemic is one that spreads over very large areas or the whole world. SRCCL's discussion of epidemics is limited, and this is presumably because, like we have just seen with the interplay between malaria and GDP, climate change is only one epidemiological factor amongst many. In discussing malaria 'where vectors are expected to increase their home range' the report points out how the causal factors are 'confounded' or entangled with other 'multiple factors', both medical and socio-economic. Mention is made of meningitis and

cardiopulmonary diseases caused by dust storms in drought-af-
flicted regions such as the Sahel. Ebola virus is singled out for
discussion because its origin was zoonotic – passed from animals
to humans – and there is *high confidence* that climate change, in
tandem with changes in land use, is a driver of human encroach-
ment into wild animal habitats. It thereby increases the possibility
of cross-species infection. That, however, is the only context in
which SRCCL ventures a confidence assessment on the matter of
epidemics.

It is worth noting that bats are often implicated in the outbreak
of new human and livestock diseases. Not only are they the world's
only flying mammal and therefore have exceptional mobility, but
they also have strong immune systems that, paradoxically, can give
rise to the mutation of more virulent pathogens that seek to con-
quer the animal's staunch resistance.[14] When this happens, infected
bats may be eaten by other wild animals. If these become part of
the bush-meat trade in 'wet' markets that deal with living animals
and unprocessed flesh, they can readily become the vectors of new
human illnesses.

The mentions of pandemics in all three of the IPCC special
reports add up to precisely zero. To have made such links would
probably have been too speculative. There are limits to what the
IPCC can incorporate as being attributable to climate change. You
could stretch a case that climate change will also increase alcohol-
ism, as that too can be linked to stresses placed on human living
circumstances. Such a link would be possible, but tenuous. Rather,
in most epidemiological matters climate change is a background
risk multiplier rather than a prime mover. Where might that leave
us in relation to COVID-19, this being the WHO's abbreviation
for 'coronavirus disease 2019'? As of March 2020, when announced
as a pandemic, it was thought most likely to have originated in bats
and probably to have jumped via an intermediary species to people.
A prime suspect is the pangolin or scaly anteater, a wild animal that
is sold as a delicacy in Chinese wet markets and the scales of which
are prized in traditional medicine. With the especially deadly SARS
coronavirus that was identified in 2003, another wildlife delicacy,
the civet cat, was the intermediary. On account of such dangerous

zoonoses, which can now sometimes be tracked by genetic finger-
printing, the Chinese authorities brought in a temporary ban on
eating and farming wild animals in February 2020.[15] For the time
being, if not beyond, nature had imposed its own animal rights.

Some commentators have tried to link zoonotic transmission of
COVID-19 to climate change. For example, Jem Bendell, who has a
large following for his blogs that anticipate climate-induced, imminent widespread social collapse, invites his readers to 'imagine' that
a climate-change related sudden crash of the insect population had
forced starving bats to migrate to find new food sources, bringing
them into contact with wild or domestic animals that carried the
coronavirus onto market stalls in Wuhan Province. He speculates:
'If the impact of Covid-19 is the latest step in the collapse of modern
societies, then it would have been another climate-driven step in
that collapse.'[16] However, to make such a theory stack up, it would
first need to be shown that flying insect populations had actually
gone through a sudden cataclysm and not, perhaps, increased, as
with British moth biomass. It would need to be shown that the bats
had survived and migrated in response, and that climate change,
rather than unrelated factors, was the smoking gun in such a chain of
transmission.

A more straightforward theory is that the virus spread probably
from bats to an intermediary host, and from there to humans, as
has happened many other times in history when people encroach
on nature. The wild meat trade, an absence of animal welfare
standards, high human population density and rapid transmission
along aviation flight paths give a more parsimonious account of
the pandemic. Climate change may be an underlying risk multiplier. It may amplify existing factors. But it is more the common
factor of lifestyles and social structures based on the mobility
that fossil fuels make possible that the finger might be pointed
towards. Pandemics like the so-called Spanish Flu of 1919 killed
between 20 million and 40 million people, this being up to
double the casualty incidence of the Great War and happening well
before climate change kicked in seriously. With or without global
warming, pandemics are always on the horizon. Episodes of 'the
plague', in whatever microbial form it might take, have always been

with humankind, especially when the combination of population and consumption pressures becomes intense. In relation to influenza, the WHO's website has long warned that 'experts agree that another pandemic is likely to happen but are unable to say when'. It adds that while the specific characteristics cannot be predicted, even conservative estimates could lead to millions of deaths very quickly.[17] Climate change may be part of that equation, but if so, and if such a claim is to be counted as real science, there must be a trail of evidence.

Some climate advocates have expressed the hope that COVID-19 has a silver lining. Here is our long-awaited wake-up call that exposes the fragility of the global economy, the corporate elite and predatory banking. National lockdowns prove that we can live without flying, without luxuries and that, if they want to, governments can act like 'big government' does in times of war. The virus, writes Brad Zarnett, a Canadian sustainability strategist and blogger, cut through all the stuttering talk of technofix, green business, reusable coffee mugs, going vegan and even Greta Thunberg's pleadings. 'Tragic but effective?' he asked, wondering if the disease could be 'the trigger that we've needed' to get greenhouse gas emissions under control. The path towards a stable climate 'will need to run through some kind of economic collapse'. The virus might be the bartender's last call from the saloon. Here's nature saying: 'Our approach will achieve success by shutting down something even more fundamental to the economy than money or oil – people.' And 'like a heat-seeking missile,' he concludes, 'COVID-19 gets directly to the source of the problem – our economic system – and it begins to dismantle it.'[18]

It was rousing stuff. It had the virtue of telling some ecological home truths. And yet, like so many latter-day jeremiads, Zarnett named all the problems but, in this piece of writing at least, proposed no practical solutions. Rather, he seemed to presume a passiveness in response to adversity, as if all else will remain equal. In fact, human ingenuity whether successful or not always seeks ways out of fixes, and so not all else remains equal. Most of us wouldn't be here on the planet today if all else had always been equal. As a reader, I was left with the impression that he sees social collapse itself as the way ahead. That may be true. If so, however, no sense came across as to what that might mean for ordinary people, perhaps already living hard-pressed lives. Neither is there any acknowledgement that the gravity of catastrophic impacts might vary with one's vantage point of privilege.

Other climate advocates have written more cautiously. Alex Trembath and Seaver Wang of the Breakthrough Institute argue that the response of world governments to COVID-19 is no model for climate change. As might be expected from an institute that champions technical and policy solutions, neither do they have much time for the climate activist and Renaissance literature scholar Genevieve Guenther, who tweeted: 'I don't want to hear one fucking word about how decarbonizing quickly enough to maintain a liveable planet is "unrealistic".'[19] They point out that even during China's shutdown – and roughly half of the country was in lockdown – emissions stubbornly remained at three-quarters of their usual level, and that a suppressed economy can quickly rebound.

Sure enough, before the month of March was out, China was looking at relaxing its environmental standards to help the car industry, and Reuters reported that the country's industry 'has gone from zero to sixty' in offering discounts 'to woo back lockdown-weary consumers'.[20] Meanwhile, the Trump administration announced that its Environmental Protection Agency would help industry by suspending its enforcement of environmental laws, a measure that will have disproportional adverse impacts on the poor and people of colour.[21] If the pandemic response is held up as a model for climate action, conclude Trembath and Wang, 'we should not be surprised if public support is less than enthusiastic.' The pandemic and climate change are not comparable. While the solutions for controlling the virus are simple and short term, those for getting to grips with climate change comprise a deeply entangled 'wicked problem': one for which there are no short, sharp shock fixes sitting on the horizon.[22]

António Guterres, the UN secretary general, took the same line when releasing a report stating that global climate action was 'way off track'. He said that although emissions had been temporarily curtailed, 'we will not fight climate change with a virus'. Whilst

the pandemic will be temporary, 'climate change has been a phenomenon for many years, and will remain with us for decades and
require constant action'.[23] In other words, the one is symptomatic
while the other is systemic.

Where, then, does this leave COVID-19 in relation to climate
change? Both share common causes linked to high population
density, a consumerist attitude towards nature and ways of life that
allow problems rapidly to propagate pathogens along flight paths.
Both are linked to the velocity and intensity of life that is permitted
by energy-dense fossil fuels. The virus will hopefully restore respect
for science and expertise, given the sudden burst in public interest
and political awakening as to why it matters. It will hopefully also
raise consciousness of our vulnerability to just-in-time global supply chains. These have been constructed to maximise competitive
economic efficiency, but at the expense of the social resilience that
can buffer knocks. It will probably enhance people's appreciation
of the need to restore a well-integrated local economic base, the
linkages and multipliers of which are good not just for conviviality, but for basic security. Later we will touch on how this might
factor in to the economics of so-called green new deals. Overall,
harrowing though it is, death may deepen our appreciation of life
and all that it gives. It may raise awareness of the need to elect
serious rather than celebrity politicians, and to engage with politics
as more than just a sport.

In all of these, as I suggested in my introduction: never let a good
crisis go to waste! If approached wisely, a crisis can be a wake-up
call. Certainly, we should not hold our breath that such lessons as
these will be learned by all. And the bottom line? (Because I love
bottom lines.) While the one is not obviously caused by the other,
the corona crisis sits enfolded in the wider climate crisis.

Food Insecurity and Cascading Risks

We use between a quarter and a third of the land's primary productive capacity for food, feed, fibre, timber and energy. Such
intensity causes multiple stresses. SRCCL has *medium confidence*
that a quarter of the world's ice-free land is undergoing degradation

– only medium, because historical baselines can be lacking, and as we've just seen, some parts of the world are greening as rainfall patterns (and also, population pressures) shift. Agricultural soils under conventional tillage are eroding at more than a hundred times the rate of new soil formation. This is all too visible if passing through an area with harrowed fields, where light soils drifted by the wind splay out across country lanes, and flush down rivers out to sea.

In contrast, the best soils are rich in humus, sticky without being claggy and therefore nicely friable – crumbly, so that roots can penetrate. Humus is mainly made up of decaying or decayed vegetative matter. You can smell the earthy richness – a real farmer with real soil will pick it up, roll it around between palm and fingers like a good malt whisky warming in the glass, and not just smell but even taste it. Not only is humus high in carbon as naturally sequestered CO_2, but it also functions as a sponge. It regulates the give and take of flood and drought. What we think of as a soil is therefore more than dust and dirt. The ground beneath our feet is this earth's living skin. It sustains terrestrial life in so many ways, which is why the gold standard of organic farming in Britain is called the Soil Association. We too, like the earthworm, are organisms of the soil.

Climate change renders latitudes closer to the equator especially vulnerable to reduced crop yields. Causes of crop failure include drought, extreme weather events such as floods, heat waves and hailstorms, and shifting ranges of what can grow where. SRCCL has *high confidence* that yields from maize and wheat have already declined in lower latitudes, and animal productivity has declined on pastoral systems in Africa. Food insecurity from crop failures pose 'cascading risks with impacts on multiple systems and sectors'. By nature, these knock-on interactions are hard to model, and the literature around them – especially at warming levels of 1.5°C and 2°C as distinct from what can currently be documented – is limited, according to the IPCC's SR1.5 report.

By an arrant twist of fate, the yields of some crops have increased in higher latitudes. Ironically, because the richer countries are, for the majority, those that most cause greenhouse gas emissions. In temperate agricultural zones a bit of extra warmth can speed germination and growth, it can extend the growing season and bring new territories into productivity. Countries like Russia or Canada might thereby gain some benefits from climate change.

A few years ago, I was speaking about land reform on Prince Edward Island in Canada. On the morning radio news programme they asked what I thought about political pressure that the island was coming under to lift a longstanding restriction on foreign land sales. The Chinese especially were taking an interest. The day before, my hosts had taken me to salute the statue of the nineteenth-century land reformer John MacKintosh, who, because he spoke Gaelic, was able to advise poor tenant farmers who were mostly of a Highland Scots descent. This eventually resulted, as his national biography so pleasingly puts it, in 'removing the curse of land-lordism'.[24] I told the radio anchor that deregulation threatened my own namesake's legacy with undoing. Just take a look at the globe. Ask what's going to happen as the Arctic melts and new shipping passages open up. 'You'll be the back garden for China's kitchen.'

On which culinary matters, with *high confidence* SRCCL holds that moving to a more vegetarian or even to a vegan diet has the potential to cut emissions massively, by more than 6 billion tons of CO_2eq a year. It's not just methane burps from cows. It's also taking pressure off the land, allowing nature's ecosystems to restore the blessing of their services. However, here I must confess in passing to falling short of more conscientious nutritional regimes. Being of a rural background where we raised, hunted and fished a good part of our diet as human members of the local ecosystem, I'm all for agroecology with high animal welfare, soil care and biodiversity, but not for zero farming.

Which reminds me of the time Vérène and I turned up at a friend's home in rural Ireland. There was our host atop the mower, perambulating a vast perimeter lawn in ever more maddening circles. As the petrol-puffing guzzler spluttered to a halt, I could contain my gastronomic continence no longer. I looked at Damien and said: 'Have you not considered geese, and sorting out the Christmas dinner while you're at it?'

Poverty, Conflict and Migration

As a driver of poverty, SRCCL stresses the 'highly challenging' complexity of trying to assess the role of climate change and related land degradation, variously described as a 'risk multiplier', 'threat multiplier' and 'stress multiplier'. While there is *very high confidence* that such multipliers are at work, there is *low confidence* around what can be concluded of the part of climate change relative to other drivers. The evidence base is surprisingly weak. How can that be so? Because 'multiple social, political, cultural and economic factors, such as markets, technology, inequality [and] population growth' all interact in compounding and confounding ways. These too shape how social and ecological systems respond.

Such murkiness extends into the fog of war. Military strategists also see climate change as a non-traditional security challenge. To the Pentagon it is a 'threat multiplier', and that, 'because it has the potential to exacerbate many of the challenges we are dealing with today – from infectious disease to terrorism'.[25] But a multiplier can be one amongst many, and is not necessarily a prime mover. SRCCL plays its hand cautiously. It cites persistent drought in Morocco in the early 1980s as an open and closed case that drove food riots and economic collapse. Likewise, the impact of the drought on livestock in Somalia driving conflict in the Horn of Africa. Similarly so in northern Mali. But some other celebrated cases are ambiguous. Whereas early studies of the conflicts in Rwanda and Sudan pointed a finger at climate change contributing to land degradation, more recent work has questioned such attribution. It favours ethnic and political prime movers.

Syria is a good example of confounded variables – and I use 'confounded' in the statistician's sense of muddled-up causes and effects. SRCCL notes that the drought preceding the unrest of 2007–10 was the longest and the most intense of the past 900 years. On the one hand, agricultural collapse displaced 300,000 families to the cities. On the other, the political backdrop was so turbulent that the IPCC finds the literature to be 'without agreement as to the role played by climate in subsequent migration'. To give just one example from such literature, the Syrian government had withdrawn fuel subsidies for irrigation pumps and failed to implement a national drought strategy approved in 2006. These and other political factors lead some experts to see the transportation of water, and not just its shortage, as having been a contributor to the crisis.[26] Such low levels of expert agreement forces SRCCL to conclude that studies of 'the risks of conflict have yielded contradictory results and it remains largely unclear whether land degradation resulting from climate change leads to conflict or cooperation'. This, it frankly concedes, is 'a major knowledge gap'.

Further afield, a raft of studies from all around the world suggest that changing rainfall patterns can drive migration, with seasonal migration often being an adaptation strategy. There is only *low evidence* 'on the causal linkages between climate change, land degradation processes (other than desertification) and migration'. Sometimes climate change becomes a social scapegoat. For example, in the Maldives there is a collective perception that blames it for migration, but the study of individual perceptions 'give more credence to other cultural, religious, economic or social factors'.

A counter-intuitive case study from Latin America is the Pacific plain and the Andean highlands. Here, the driver of migration is not land degradation, but high land quality that opens new horizons from the revenues of agricultural produce. For many young people raised in rural areas, cities can represent freedom and the chance to make both money and a name. Another example that might easily be misread with a privileged Western disposition is from Bangladesh, where it is mainly men who migrate under the pressure of heat stress, 'leaving women and children to cope with increasing effects of natural disasters'. Whether this is selfish, or a desperate family adaptation measure to remit money back home, is a question that would require further study to establish.

As I looked at these sections of SRCCL, I could not help noticing the gap between the simple perceptions that a reasonably well-informed person in the Western world might have, and the complex realities on the ground elsewhere. We live in a hyperconnected age where events that would once have gone unreported immediately appear in sound bites and video clips on our screens. If one shocking event somewhere becomes a media meme – say, floods or wildfires – reporters elsewhere will be alert for them to feed a receptive audience until boredom with sameness sets in. The fact that our psychology has evolved to be on the alert for threats might be why most of us find bad news more compelling than good. Media, driven by the ratings and commercial demands, will provide its consumers with what gets the clicks. We heard much about the Californian wildfires and power outages in the autumn of 2019 – caused both by drought and a lack of maintenance by the bankrupt utility PG&E that could have prevented power-line sparking. But how many of us heard that, at the same time, the Arizonan desert was being watered by the tail end of Caribbean hurricanes which brought in very welcome record rainfall?[27]

We count the curses but not the blessings, although with climate change the upsides and the downsides are vastly asymmetrical. But note another thing in passing. Even in the examples just given, my bias in that last paragraph has been towards parts of the world that are well reported, well funded, well researched, 'white' and English-speaking in their dominant cultures. A storm will be reported across the world from the American Midwest that wouldn't cause a Hebridean sheep to lift its head. And there we go again – pitching it from within my own worldview – but, if you please, a hardy blackface ewe, none of those new-fangled breeds.

Called in by the Ambassador

Levity apart – a compensatory levity – I felt enormously frustrated the more I examined the IPCC's take on poverty, conflict and migration. Attempting to triangulate this by reading up around it from several different angles didn't help either. The IPCC's caution seemed accurately to reflect a diversity of opinion in the professional literature. The activist in me wanted to blast through such seemingly pedestrian caution. To shout out, 'Of course climate change drives these evils!' I found out, however, that relief workers in the field have learned the hard way that to misattribute risks, impacts and their causes results in interventions that can be counterproductive or ineffective. If we don't like evidence-based science, whether in the physical or social domains, with what might we better it? That's always the question to those who say that science is intrinsically biased towards science. They might answer, 'My feelings!' Well, OK, but then how do we decide between contesting feelings at a social policy level? The quest for objectivity can be indeed a values-vacant vice. But if administered within democracy, it can also be a virtue.

The above duly acknowledged, as I completed working on these themes I happened to speak at an event on communities, land reform and nature conservation that had been organised in Edinburgh by the John Muir Trust. The chair was Hugh Salvesen, a retired career diplomat of nearly thirty years standing, latterly the British ambassador to Uruguay. With a well-honed homing instinct, he asked me what I'd been finding out about migration from the IPCC's reports. As he sees it, migration from Central America is a major up-and-coming issue, same with migration from North Africa to Europe. I dutifully summarised what's been said above.

'Yes,' he said, shaking his head, 'but that understates migration as a political driver.'

And I thought: yes, the ambassador's got a point. There's something out of kilter in the political psyche about migration. The UN's 2018 World Migration Report points out that 3.3 per cent of the world's population are migrants. This is up from 2.3 per cent in 1970. It emphasises that 'remaining within one's country of birth overwhelmingly remains the norm'. It almost seems to signal a calming of concern. Then I went back and read the foreword by the agency's director-general. He warned about the consequences of being in an era of 'post-truth politics', of 'information overload' and of 'fake news'. The chance of really understanding migration from a solid evidence base has diminished, he said, 'at a time when, more broadly throughout the world, facts and expertise seemed to have increasingly taken a back seat to opinion and politics'.[28]

I found myself thinking of the 'Breaking Point' poster that the UK Independence Party put out during the Brexit campaign. It was reported to the police for inciting racial hatred by the way that it depicted a 'swarm' of displaced dark-skinned men. In a widely used photograph, the politician Nigel Farage stands before it pointing, as if to say: 'Watch out! They're coming over here.'[29] And I thought, that's what the director-general was concerned about. That's what the ambassador meant. The vulnerable can so easily be used as scapegoats onto which the unresolved problems of a society, and of individuals within it, can be projected. To our list of climate threat multipliers, we might add hard-line politics.

Framing Climate and the Twin Drivers

The IPCC's special reports are not leisure reading. Working through them can feel like ploughing through a soil compacted by compromises, concessions, caveats and counterpoints. In his book on the psychology of climate action, Per Espen Stoknes looks at the language in which the reports are written. Scientists will want to discuss uncertainties, but the public want certainties. He takes as an example the Fifth Assessment Report in 2014. It spoke of '*medium confidence* before and *high confidence* after 1951' that, 'averaged over the mid latitude land areas of the Northern Hemisphere, precipitation has increased since 1901'. He goes on to ask: 'After having read this, can I tell my buddies that it rains more now than before?'[30] While the language is precise, it would hardly win awards from the Plain English Campaign.

Behind closed doors, thrashing out such language leaves blood trails on the carpet. For example, SR1.5 struggled to get through governmental acceptance. Saudi Arabia led a group of oil-producing nations over a procedural technicality, claiming that there was inadequate research to distinguish between the effects of 1.5°C and 2°C of warming.[31] The heroes of the day are not just the scientists committed to what can be a costly objectivity, but the so-called 'faceless' diplomats of many nations, who navigate the ark as best they can through sharks and shoals in what can be the shallow waters of partisan interests.

A wider issue than language is the framing of the debate. To explore the effects of different societal choices on emissions and temperatures up to 2100, SRCCL models five scenarios called Shared Socioeconomic Pathways (SSPs).[32] For example, SSP1, with low challenges to mitigation and adaptation, envisages a move towards sustainable lifeways, world population falling to 7 billion, high incomes with fewer inequalities, less resource-intensive consumption, eco-friendly lifestyles and technologies, and – presumably to keep everybody in the boat – free trade. In contrast, SSP3, with high challenges to mitigation and adaptation, envisages political turbulence and insecurity, world population rising to 13 billion, low and very unequal incomes, resource-intensive consumption, competitive lifestyles with slow technological change and – presumably to stop anybody from jumping ship – barriers to free trade.

This set of scenarios – one of several used at different times and in different contexts by the IPCC – are tuned to the issues under exploration in SRCCL. Of particular importance is that they show the implications of differing levels of both population and material consumption. Notwithstanding that, SRCCL, like other IPCC reports, makes only passing mention of these two key factors, and it adopts, without comment, a UN projection that population will reach 9.8 billion by 2050 and 11.2 billion by 2100. Throughout its 1,542 pages, family planning is mentioned just once, in passing, in parentheses, and buried in the woodwork of a discussion about multi-level policy instruments available to governments. There are but five passing mentions of 'over-consumption', the strongest of which is merely to express *high confidence* that 'changes in consumer behaviour, such as reducing the over-consumption of food and energy, would reduce greenhouse gas emissions from land'.

But where is any deeper sense of exploring these as policy instruments to tackle the most pressing global problem of our time? One climate scientist said to me privately: 'Questioning whether the current state of affairs or trends are good or not strays too far out of the safe quasi-apolitical zone that the IPCC strains for to maintain its existence.' Accordingly, I think of population and consumption as the *twin drivers*. As we will see in Chapter 7, there are reasons historically why the debate around these has been framed in such a restrictive manner. Some of these reasons, especially where population policy is concerned, have the best intentions. But there is a downside.

The downside is that it easily becomes normalised to think that there is no alternative to the 1.5°C or 2°C rise – or even the 3°C rise by 2100, to which some voices are now trying to lower aspirations – and that this is to be rolled on towards relentlessly in a technocratic future. We then end up with the spectacle as seen at the World Economic Forum at Davos in 2020. Here Steven Mnuchin, the US treasury secretary, rejected carbon taxes and with patrician pizazz argued that '. . . the costs are going to be much lower 10 years from now, because of technology'.[33] Talk of NIMTOO – not in my term of office. As for tackling consumerism, tally-ho, we'll not think about that one, because what sane politician could possibly apply the brakes?

The 2030 Agenda for Sustainable Development

That criticism made and the IPCC's framing limitations recognised, a bright spirit nevertheless shines through SRCCL. It holds out *high confidence* that mutually supportive climate and land polices can potentially 'save resources, amplify social resilience, support ecological restoration, and foster engagement and collaboration between mutual stakeholders'. It brings out the importance of women's land rights and their land management knowledge, as well as *high confidence* that indigenous and local knowledge can contribute to increasing food security, conserving biodiversity and combating land degradation. Such resilience will be furthered through approaches that enhance participation, social learning, citizen science and shared action on land tenure, including holding land collectively. Neither are such aspirations made in a policy vacuum. What draws them together, what gives them legs and legitimacy, is that in common with SR1.5, SRCCL berths itself firmly alongside the integral human development goals of the UN's 2030 Agenda for Sustainable Development.

The term 'sustainable development' was defined in *Our Common Future*, the UN's seminal 'Brundtland Report' of the World Commission on Environment and Development of 1987, as being: 'Development that meets the needs of the present, without compromising the ability of future generations to meet their own needs.'³⁴ The 2030 Agenda was adopted by UN Resolution in 2015. Its seventeen integrated Global Goals include climate action, eliminating hunger and poverty, responsible consumption and production, quality education, health and wellbeing, reduced inequalities, and peace and justice. SRCCL holds *high confidence* that most responses to climate mitigation and adaptation can 'contribute positively to sustainable development and other societal goals . . . and have the potential to provide multiple co-benefits'.

Climate action cannot be a piecemeal or a partisan affair. All stakeholders must be participants. Although 'social justice' is mentioned only four times, once as part of a definition of sustainable development, a picture emerges of the kind of world that is needed to carry humanity forwards with dignity. In a headline statement, SRCCL's 107 authors express *high confidence* that adaption and mitigation measures, if wisely and participatively applied, can, 'for those who are most vulnerable', bring a range of social, ecological and economic co-benefits that 'can contribute to poverty eradication and more resilient livelihoods'. The concept of resilience is inherent to all of this. SRCCL describes it as:

> The capacity of interconnected social, economic and ecological systems, such as farming systems, to absorb disturbance (e.g., drought, conflict, market collapse), and respond or reorganise, to maintain their essential function, identity and structure. Resilience can be described as 'coping capacity' . . . viewing the land as a component of an interlinked social-ecological system; identifying key relationships that determine system function and vulnerabilities [and] identifying thresholds or tipping points beyond which the system transitions to an undesirable state.

One can indeed be sceptical about the UN's 2030 Global Goals. Perhaps in a bid to keep the Davos set in the boat, the eighth goal pitches for what it calls 'sustained, inclusive and sustainable economic growth'. There's an oxymoron if ever there was one, unless the growth anticipated is non-material. Although, as we have just noted, SRCCL mentions 'social justice' directly only four

times, a search within its text for 'free trade' comes up with exactly
the same number of occurrences. For rune readers, a message might
be read there in terms of how consensus is built up in the writing
of such documents, where conflicting underlying interests might
be at stake.

The language of 'sustainable development' can convey both
hope and vision for life on earth. But it can also be reduced to a
mindnumbing morass of bureaucratic platitudes that violate the
inner spirit. This is why, rather than abandon the term to the rav-
ages of the 'sustainable growth' merchants as some would do, we
must reclaim it.

Properly used, 'development' comes from the Old French, and
behind that, possibly from an older Celtic or Germanic root.[35]
It means to unwrap or unfold. The same root gives us 'envelope'.
It can be helpful, therefore, to think of true development as being
to unfold what is enveloped. To enable something, or a person or
community, to realise their full potential from within. To do so in
ways that have resilience and can endure. It follows that development that is endogenous – having its genesis from within – will
be authentic, which is to say, self-authored. Development that is
exogenous – imposed from outwith – will be, in the absence of
what the UN calls 'free, prior and informed consent',[36] yet another
form of violence, yet another face of colonisation.

The American ecologist Aldo Leopold wrote of what he called
'the land ethic'. His words add depth and focus to development
that is sustainable. 'A thing is right,' he said, 'when it tends to preserve the integrity, stability, and beauty of the biotic community. It
is wrong when it tends otherwise.'[37] As we will see later, that which
is right involves an ever-deepening understanding of relationships,
a cause for joy.

CONTAINING GLOBAL WARMING TO WITHIN 1.5°C

The benchmark for climate action targets through this century is the Paris Agreement, being the outcome of the 21st Conference of the Parties, attended by the governments that are members of the United Nations Framework Convention on Climate Change. COP 21 took place in Paris at the end of 2015. By early 2020 it had been ratified by 189 of the 197 parties.[1] Signatories include China, Russia, India and the countries of the EU. Non-signatories are a mixed bag of high emitters and failed states. Although the US was a signatory at the time of writing, President Trump signified his intention to withdraw in November 2020 in furtherance of his America First doctrine. As he put it: 'I was elected to represent the citizens of Pittsburgh, not Paris.'[2]

'Paris', as the agreement gets known, lacks any binding enforcement mechanism. Although it has the status of a UN treaty its pledges are aspirational. Nevertheless, this standing gives it leverage in international relations. As laid down in Article 2A, the primary objective is: 'Holding the increase in the global average temperature to well below 2°C above pre-industrial levels and pursuing efforts to limit the temperature increase to 1.5°C above pre-industrial levels, recognizing that this would significantly reduce the risks and impacts of climate change.'[3]

Special Report: *Global Warming of 1.5°C*

Global Warming of 1.5°C,[4] abbreviated to SR1.5, was released by

the IPCC in October 2018 at the request of the UN's Framework
Convention to explore how to fulfil the Paris Agreement. It examined around 6,000 recent scientific studies, was compiled by more
than 200 authors from all over the world and scrutinised by some
1,100 peer reviewers. In generating media reports and headlines
internationally, it bolstered the work of scientists, policy officials
and concerned politicians, and helped fuel a rising new wave of
climate activism that spanned the 'Fridays for Future' school
climate strikes associated with Greta Thunberg, the demands of
Extinction Rebellion and many other endeavours less prominent
in the public eye.

The IPCC's remit was: '. . . to provide a Special Report in 2018
on the impacts of global warming of 1.5°C above pre-industrial
levels and related global greenhouse gas emissions pathways.' In
conceding the possibility – most would say a foregone conclusion
– of a target overshoot approaching 2°C, SR1.5 includes models
showing how temperatures might be brought back down to 1.5°C
again by more stringent measures taking effect later in the century.
It therefore set itself a dual target, whereby:[5]

> In model pathways with no or limited overshoot of 1.5°C,
> global net anthropogenic CO_2 emissions decline by about
> 45% from 2010 levels by 2030, reaching net zero around
> 2050.

> For limiting global warming to below 2°C, CO_2 emis-
> sions are projected to decline by about 25% by 2030 in
> most model pathways and reach net zero around 2070.

The starting point of SR1.5 was the baseline that the world had
already warmed by 1°C by 2017. It anticipates that, in the absence
of rapid and effective action, the combination of existing carbon
concentrations and new emissions added all the time are *likely* to
raise world temperatures to 1.5°C above pre-industrial levels some-
time between 2030 and 2052. By 'net zero', it means that any further
CO_2 emissions in one context are offset by an equal compensating
level of removal in another. An example is offsetting the carbon

cost of a flight by paying to plant trees to soak up CO_2 – a practice based on the questionable assumption that any trees planted will be cared for, and never burnt or felled and allowed to decompose back to atmospheric carbon gases.

Why did the Paris Agreement choose 1.5°C as a relatively safe threshold? One answer found in SR1.5 is that here the earth will have reached its 'limits to adaptation and adaptive capacity for some human and natural systems', and that if the warming over-shoots towards 2°C, then with *high confidence* some ecosystems will be irreversibly damaged or lost. A harbinger of such an occurrence would be the Australian forest fires of 2019–20. On preliminary estimates, these destroyed half of the remaining 40 million-year-old Gondwana rainforest in such regions as the Blue Mountains and the world heritage Nightcap National Park, including parts that had never before been logged or burned.[6]

If temperatures can be constrained or brought rapidly back down to 1.5°C, then the projections for global sea level rise for the period spanning the turn of the twentieth century to the end of the twenty-first should be within 0.26–0.77 metres. An overshoot to 2°C raises this to 0.35–0.93 metres. As the mid-points of these ranges are 0.51 and 0.64 metres respectively, one way of looking at it in round figure terms is that the difference between 1.5°C and 2°C is half a metre and two-thirds of a metre by the century's end, which is to say within a single human lifetime. On top of that, SR1.5 reminds us that polar ice sheet instability is more likely to be triggered above 1.5°C.

It is all very well to say 'build walls and raise everything up', but consider just one small example, again from a well-documented part of the world from where these types of stories more readily emerge. Officials in the Florida Keys, a string of islands linked by causeways off the American coast, have found that to keep a threatened three-mile stretch of road dry in 2045 would mean ele-vating it by 2.2 ft – that's two-thirds of a metre. This would cost $128 million, and some 300 miles of similar stretches of road also need raising. Multiply that up for a crude estimate, and you hit $12.8 billion. The county can't afford it, the taxpayers won't bear it and the mayor fears a string of lawsuits from irate householders who claim the law entitles them to have such lifeline links maintained.

'What is government for? They're supposed to protect your property,' insisted one resident.

'Maybe we should think about stopping, or trying to stop, the cause of the water rising,' said another. 'At what point will the road be high enough?'[7]

'We've Only Twelve, Eleven, Ten . . . Years Left'

To turn that round another way, at what point does the world run out of 'carbon budget' to keep within the 1.5°C threshold? From where did climate change derive the slogan, as it was when SR1.5 came out, 'We've only got twelve years left,' and counting down progressively since then, as if the next space shot is heading off to colonise Mars courtesy of Elon Musk?

The twelve-year countdown is never mentioned by the IPCC, but the calculus based on SR1.5 looks like this. Since the industrial age began and through until the end of 2017, anthropogenic emissions have accumulated 2,200 billion tons (or gigatons) of CO_2 in the atmosphere. To have *medium confidence* of a 50:50 chance of containing global warming to within 1.5°C of pre-industrial temperatures, only an estimated 580 billion tons of capacity remain. If we want to raise the confidence level to a 66 per cent chance, we should think of it as only 420 billion tons remaining of the so-called 'carbon budget'.

How long will that last us? Well, annual CO_2 emissions for 2017 were running at 42 billion tons. Average out those two budget figures, divide by forty-two, and there's your twelve years, starting in 2018 and lasting through to 2030. I'm guessing that's how, in rough-and-ready terms, somebody somewhere came up with that date and caption. Reasonably enough, it caught on. What's unhelpful is that we've had a string of 'Only so-many years left . . .' warnings going back to the 1990s, the vagueness and inflexibility of which left hostages to fortune. But what's helpful this time is that it's grounded in measurable science. It passes a key, if sometimes glib, acid test of scientific method: 'If you can't count it, it doesn't count.' And for campaigning purposes, that brings to life the urgency of the primary Paris target. It shows why it matters that we cut emissions first by 45 per cent by 2030, and then on down to net zero by 2050.

But how? What does it mean the governments of the world must do? And what in a context where SR1.5 states that the national mitigation ambitions that have so far been submitted to the UN under the Paris Agreement 'would not limit global warming to 1.5°C'? In fact, with *medium confidence*, they point more towards 3°C by 2100. In other words, we're already failing by a factor of at least 100 per cent. To get to grips with what needs doing, SR1.5 maps out four Illustrative Model Pathways for differing mitigation action strategies. Here is yet another set of models, overlapping with, but not identical to, the two sets of models that we have previously touched on – the RCPs and the SSPs. One can almost sense the fun draining from the hard realities as SR1.5 avoids the acronym IMP for its Illustrative Model Pathways, naming them just pathways P1, P2, P3 and P4. Here we need not enter into detail. Suffice to say that each explores a different CO_2 mitigation option showing how, through to 2100, it would be technically possible in principle to limit global warming to 1.5°C with no or limited overshoot up to 2°C.

The framing here is crucial to understand. In order to work within what could make Paris possible, all four pathways assume a commitment to low emissions scenarios as their baseline. Two assumptions here are worth highlighting. Quantitatively, that there will be investment in the world energy system of around $2.4 trillion between 2016 and 2035 – that is, about 2.5 per cent of global GDP. And qualitatively, a recognition that 'social justice and equity are core aspects of climate-resilient development pathways' that could slow warming to 1.5°C, and this so that the 'inevitable trade-offs' of decarbonisation can be undertaken 'without making the poor and disadvantaged worse off'.

Within that hopeful framing, the pathways vary in their assumptions about levels of technological innovation, decarbonisation of the energy supply, downsizing energy demand, the degree of international cooperation and social shifts towards sustainable and healthy consumption patterns. All four scenarios would entail 'rapid and far-reaching transitions' in how we live. All require a global transformation of energy production, transport, agriculture, the built environment and industrial processes. Not for nothing does SR1.5 invoke the word 'unprecedented' twenty-eight times.

Nuclear and Geoengineering

If net zero and 1.5°C are to be achieved, even with a temporary overshoot to 2°C followed by sharp remediation later, most of the world's energy supply must be decarbonised and what can't be must be offset. Fossil fuels must be replaced with renewables such as wind, wave, tidal, solar, hydro and biofuels grown from trees and other plants, as well as the rapid development of energy storage technologies and 'smart' means of regulating demand to fluctuating supplies. However, and in addition to all these, on most 1.5°C pathways, the share of nuclear energy is projected to increase. This, either with existing reactor technologies, or bringing in generation III/IV reactors that are still in development. Nuclear currently provides about 10 per cent of the world's electricity and exceeds the contribution of renewables.[8] If it looks set to increase, where stands the debate that builds it into climate pathways?

Advanced reactors are said to have high intrinsic safety specifications, being capable of shutting down without operator intervention. Some designs can consume waste fuel from older reactors or breed further fuel, thus reducing both disposal issues and the need for mining. Neither are they only on a massive scale. In 1986 Drax power station in Yorkshire had a coal-fired capacity of 4,000 megawatts and the Hinkley Point C nuclear station, currently under construction complete with nearly a kilometre of protective sea wall, has a capacity of 3,200 megawatts.

In contrast, China expects to have a tiny 125 megawatt 'small modular reactor' in operation by 2025. Nuclear submarines already have them, packed into the back of boats less than a hundred metres long. Westinghouse in the USA expect to have their *eVinci* 'micro-reactor' producing as little as 200 kilowatts – 200 bars of an electric fire – though typically, more like 1–25 megawatts and thereby offering much more flexibility of usage. With the technology expected to be tested by 2022, moving to commercial production in 2025, these will be small enough to fit onto the back of a truck. They claim 'walkaway inherent safety', have no external coolants or pumps and therefore no need for risky positioning by the sea, provide up to ten years running without refuelling, claim low proliferation risks due to encapsulation of the fuel – and zero carbon operating emissions. Any downsides seem to have escaped the advertising team. A widely published graphic claims, 'Fossil fuels can't compete' and 'Readily compatible with renewables'. It shows an installation at a wind and solar farm, the implication being that when the wind or sun go down, instead of needing storage the reactor will kick in.[9]

SR1.5 acknowledges 'the political processes triggered by societal concerns' – including constraints of cost, risk and the public perception of risk – that render nuclear a limited or unacceptable option in many countries. It is not for the IPCC to say whether such a technology is right or wrong. Their part is to advise on what is technically feasible, and to build existing and emergent realities into their models. But what about the rights and wrongs? The arguments against nuclear power over the past three-quarters of a century have been multi-levelled. Perhaps the strongest has been the definition that an unacceptable risk is an uninsurable risk. As with many environmental questions – GM crops are another example – the most challenging question to an industry lobby is very simple: 'Who is your insurer?' But today, carbon has leapt up the league of actuarial nightmares. The devil's own dilemma shows in Germany, which has opted to phase out its remaining nuclear plants. Now it plans to work existing pits of filthy lignite coal to exhaustion, and even open up a new one. As of 2016 its power stations each year emitted nearly a third of a billion tons, about 1 per cent of world fossil fuel CO_2 emissions. Its electricity had a carbon footprint of 516 grams per unit – nine times that of nuclear France at merely 58 grams.[10] If we want to know why Eurostar claims such incredible carbon efficiency for its Channel Tunnel trains, just ask from which end of the pipe it draws its electricity.

Where all this leaves the balance of arguments – from costs, to safety, to issues of power and control, to military proliferation, to waste disposal – goes beyond my competence to judge. The same goes for hopes held for nuclear fusion (as distinct from conventional fission) and its dream of virtually endless, relatively 'clean' energy that has, since the 1960s, been forecast as 'only thirty years away'. The backdrop is that many environmentalists of my generation, children of the Hiroshima, Nagasaki and Cuban Missile Crisis age, were chilled to the marrow by Windscale, Three Mile Island and Chernobyl. Much energy and many markers of personal identity within the green movement centred around the sunshine logo, 'Nuclear power, no thanks!' With some discomfort, I would propose a question that without foregone conclusions might lead to a fresh evaluation. Given what we now know of the threat from carbon: which source of energy today has gone most critical?

While in the camp of high technology, what about geoengineering as a way of cooling planet earth? What about ideas like spraying sulphates into the upper atmosphere to mimic the haze of major volcanic eruptions, or placing giant mirrors in the sky to reflect some of the sunlight back to outer space? SR1.5 calls such ideas 'solar radiation modification'. It says that such technologies are not included in any of the available mitigation pathways that it has assessed. They would offer no respite to ocean acidification. The shifts in weather patterns would be very difficult to predict, with benefits to one part of the world perhaps weighing against another. In all, there are 'substantial risks . . . related to governance, ethics, and impacts on sustainable development'. These large uncertainties and gaps in knowledge 'constrain the ability to implement solar radiation modification in the near future'.

Moreover, if once started, solar radiation modification would have to be maintained indefinitely. To switch it off, or be knocked out, would cause immediate 'termination shock'. That's the jargon that the IPCC uses. It means hot turkey on a junkie planet.

The Case for CO_2 Removal

We are left with the problem that greening the energy supply, and even escalating nuclear, is not going to be enough to hit the Paris targets. We saw that anthropogenic emissions have already put well over 2,000 billion tons of CO_2 into the atmosphere, and escalating.

SR1.5 makes the chilling finding that even deep cuts to emissions are not going to be enough. As such, and with *high confidence*: 'All pathways that limit global warming to 1.5°C with limited or no overshoot project the use of carbon dioxide removal (CDR) in the order of 100–1,000 billion tons of CO_2 over the 21st century.' This, in the face of *high confidence* of running into 'multiple feasibility and sustainability constraints'.

The challenge posed by CDR lies in the fact that CO_2 makes up only 0.04 per cent of the atmosphere, the remaining 99.96 per cent is mainly nitrogen and oxygen. As an end-of-pipe approach, it is like sucking needles out of haystacks. That said, ways have been proposed. These include 'fertilising' the oceans by sprinkling them with iron filings. If it worked – which trials suggest is doubtful – algal blooms would be provoked that would sink down to the bottom, deposit carbon into seabed sediments, and hopefully without the usual problems that algal blooms bring with them.

Another way is to crush up vast quantities of rock and spread it over the land to absorb the CO_2 through 'enhanced weathering'. Hopefully, nobody will mind the scale of quarrying, transportation and energy required, and the farmers will be friendly.

What SR1.5 posits as a 'socially more acceptable' approach than many of its CDR cousins, would be to grow a lot of trees, turn them into charcoal, and mix such carbon into agricultural soils. SR1.5 acknowledges the value of such 'biochar' for some soils and in appropriate local contexts. However, the amount of land required for tree plantations, the cost per ton of capturing CO_2 in this way, and the eventual carbon saturation of the soil forces it to conclude that, at a global scale, the potential is limited.

At the end of the day, a raft of different approaches to CDR might all contribute slices of the cake, but each has its downsides and unknowns, especially at such unprecedented scales. SR1.5 therefore narrows down its focus to what it sees as being the two most promising 'negative emissions technologies'. The first, which is a purely technical approach, is called 'direct air capture'. The second, which is a combination of a biological approach with a technical one, is called Bioenergy with Carbon Capture and Storage – fondly known to aficionados as BECCS. These next two sections are technical, and the general reader might want to skim over them to the next section, 'The Land Demand of Action'.

Direct Air Capture and Storage of CO_2

Imagine vast arrays of solar- or nuclear-powered fans out in remote locations. These would blow immense quantities of air through a chemical process, energised by more solar power and natural gas, to scrub the CO_2, recover it from a chemical solution and compress it for transport. Having achieved such 'carbon capture', durable 'storage' could hopefully be achieved by pumping it down into suitable rocks such as disused oil-and gas-field strata. Alternatively, by adding yet more energy and further chemical reactions, the CO_2 so captured could be reverse engineered back into jet fuel – and the process then repeated all over again.[11]

Such technologies are already under development by private companies, drawing on high-risk venture capital and research grants. A leading example is the Canadian start-up, Carbon Engineering, driven by the brains of a Harvard physics professor, David Keith. While acknowledging that 'it is difficult to estimate the cost of a technology', he hopes that 'depending on financial assumptions, energy costs, and the specific choice of inputs and outputs' the process can be commercially scaled up to enable atmospheric CO_2 to be captured and compressed for as little as \$94 to \$232 per ton.[12] In round figures, that might add just over 20 per cent, by way of a polluter-pays fuel tax, to pump prices in Europe. In America, the mark-up relatively speaking would seem roughly double, as their fuel has a lower baseline cost owing to very low taxation.[13] Why not just plant forests instead of plastering the land with solar voltaic panels? Keith tweets his reasoning thus: 'Plants have solar-to-fuel efficiency of under 1 per cent for practical systems. And they use over ten times more land and more water.'[14]

A joint study published in 2018 by the Royal Society and the Royal Academy of Engineering looked at CDR options. It concluded that, in technical terms, direct air capture could be ready for large-scale deployment by 2050, and it cited Keith's Carbon Engineering costings to illustrate its potential affordability.[15] However, it also said that direct air capture has two major challenges: an adequate supply of low carbon energy and, depending on the technology used, water for the scrubbing process. A similar study the same year from the European Academies' Science Advisory Council – the body that advises governments of the European Union – was less upbeat. It conceded that some carbon capture technologies have already proven themselves in demonstration projects, for example, in removing CO_2 from power station and industrial exhaust. However, the negatives of scaling things up to processing the earth's atmosphere would include enormous capital costs, scale of equipment and the vast energy requirements of capturing, compressing, transporting and pumping down the CO_2.[16]

The energy demands of direct air capture and carbon storage is the focus of a further recent paper in one journal from the *Nature* stable. We have seen that SR1.5 anticipates a need for 100–1,000 billion tons of CO_2 removal between now and 2100. Were this to be captured at a rate approaching 30 billion tons a year towards the end of the century, the energy required would exceed half of today's entire global demand, and that at a time when other energy sources might be in competition for renewable resources to decarbonise in the race to reach net zero emissions. While not dismissing such technologies, and seeing them as having a place alongside other CDR approaches, the study concludes (and some researchers think optimistically so) that: 'The risk of assuming that direct air capture with storage can be deployed at scale, and finding it to be subsequently unavailable, leads to a global temperature overshoot of up to 0.8°C.'[17] Put simply, to rely on such a technofix could leave the world up the creek with a broken paddle. As Duncan McLaren, a former director of Friends of the Earth Scotland, puts it: 'There is a huge temptation to "bank these promises" of future removals against slower mitigation now,'[18] leading him to describe them as 'technologies of prevarication'.[19]

SR1.5 gives as its bottom line the gently understated observation: '. . . only a few published pathways include CDR measures other than afforestation and BECCS'. If direct CO_2 removal was that promising, why would it not factor in more? It may just be the novelty of the technology. Sometimes, however, one wonders if gee-whiz technologies that make a media splash can be as much a pitch for grants and venture capital as realistic ways to save the world. That caution sounded, direct air removal is one to watch. The irony is that it would take colossal amounts of energy to deal with the consequences of colossal energy profligacy. The first priority is 'energy obviation' – reducing the need for using so much.

We will leave the high tech there. What then, of biological approaches?

Land Use and Bioenergy Carbon Capture and Storage

We saw that land use accounts for 13 per cent of global CO_2 emissions as well as contributions to other greenhouse gases. The wider term used in describing it is Agriculture, Forestry and Other Land Use, affectionately known as AFOLU. Moving to zero carbon requires a range of measures, some of which have the potential of not just ceasing but reversing emissions and thereby bringing about CDR. Such steps can include reducing the area of land under meat production, especially where intensive and reliant on imported feedstock; farming practices such as organic agriculture that build up soil quality by encouraging the sequestration of carbon as humus or charcoal; the conservation of peat and other wetlands that build up carbon deposits in waterlogged, acid environments that discourage decomposition; and afforestation (the planting of new forests) and reforestation or reafforestation (the regeneration of previously afforested lands), that capture carbon not just in the soil from such processes as leaf-fall, but hold and store it in the forest canopies themselves – a root and branch affair.

AFOLU at its best works *with* nature rather than short-circuiting it. Plants 'breathe' vast quantities of air. In photosynthesis, whether of microscopic green algae in the sea or giant baobab trees in Africa, plant life sequesters CO_2 and, with the energy of sunlight, turns it into sugars, starch, oils and cellulose in the form of wood or other fibres. Burning such 'biofuel' reunites the carbon with oxygen and releases CO_2. The heat of bygone summers can then be changed to electricity in power stations or used to warm homes. Alternatively, plant products can be made into a fuel gas, extracted as a product such as palm oil, or fermented into ethanol, ethyl alcohol, that, if not consumed in ways that more become the barley, can be blended with liquid fossil fuels and sold at the pump.[20]

The exhaust gases can in principle be captured, compressed and pumped underground for durable safe-keeping. Such bioenergy with carbon capture and storage partly uses nature's processes to grow the fuels, and partly uses technology to dispose of the CO_2. In contrast, wider AFOLU measures work by leaving the carbon stored in nature, albeit on the assumption that the ecosystems in question will be protected down through time. Also, as SRCCL put it: 'In all forest-based mitigation efforts, the sequestration potential will eventually saturate unless the area keeps expanding, or harvested wood is either used for long-term storage products or for carbon capture and storage.'

With *medium confidence* SR1.5 says 'some AFOLU-related CDR measures . . . could provide co-benefits such as improved biodiversity, soil quality, and local food security.' However, if deployed at scale, 'they would require governance systems to be set in place to . . . protect land carbon stocks'. In other words, there's no point in planting or regenerating forests if they fall prey to fire, grazing pressure or unsustainable felling. There's the problem on a crowded planet. It leaves the IPCC having to express *high confidence* that this approach to CDR 'may compete with other land uses and may have significant impacts on agricultural and food systems, biodiversity, and other ecosystem functions and services'.

Oil palm, grown to crush the nuts for biofuel, is a case in point. During the 1990s, vast plantations were created in Malaysia and Indonesia to satisfy the emerging European 'green' market to reduce the carbon footprint of diesel. Agroforestry consultants did a roaring trade through such bodies as the UK government's Commonwealth Development Corporation. Quite what indigenous people living on and with the affected lands felt as they lost their heritage and autonomy, and became a pool of cheap waged labour, was a question too small to block the path of 'progress'. It is, however, big enough to raise the question of the so-called 'slavery footprint' by which others might offset their carbon karma.

Today, the obvious has become accepted, that clear-felling

rainforests and dousing them with chemicals has been a tragedy. To the chagrin of countries that went down this path, largely at the behest of European nations, the EU ruled in 2019 that most palm oil could no longer count toward its renewable transport targets and ruled it out from future import duty reliefs.[21] What had once been touted as a bright idea, a green cash cow that would deliver foreign currency with 'development', fell to the law of unintended but not unforeseen consequences.

The Land Demand of Action

If CDR to the order of 100–1,000 billion tons is needed over the remainder of this century to keep warming under, or to bring it back down to, 1.5°C, what of the land requirement of which SR1.5 expresses concern? One study by a team of forty international researchers, led from the University of Aberdeen and published in *Nature Climate Change*, asked how much land it would take to grow the biofuel necessary to sequester a mere 3 billion tons of CDR per annum? They came up with 700 million hectares and point out that this is close to half of the world's current arable land and permanent crop area. This, of course, would mostly not entail the rich rewilding of restored biodiversity, but in the case of trees, vast monocultures of fast-growing plantation species to be industrially clear-felled. If bioenergy with carbon capture and storage were to be deployed at such a scale, 'there would be intense competition with food, water and conservation needs'. They concluded that deeper cuts to emissions might make more sense than trying to chase after what's already left the tailpipe.[22]

As the climate scientists Kevin Anderson and Glen Peters wrote in *Science*, 'The allure of BECCS and other negative-emission technologies stems from their promise of much-reduced political and economic challenges today, compensated by anticipated technological advances tomorrow.'[23] We heard the same concern about direct air capture. Both may be examples of 'moral hazard' or 'mitigation deterrence' – namely, the creation of a false veneer of action or safety that kicks the need for more immediate measures out into the long grass.

The IPCC's report writers were well aware of these concerns and constraints. It emphasised that the pathway models in SR1.5 were just that: 'illustrative' and 'conceptual' models to help think about the problem. Whether they would be socially and politically acceptable is a different question than what may or may not add up in purely technical terms. So it is, with *high confidence* and, one suspects, a heavy heart, that the authors of SR1.5 arrive at a humbling conclusion: 'CDR deployed at scale is unproven, and reliance on such technology is a major risk in the ability to limit warming to 1.5°C.' These words provide bleak comfort for any hopes of easy remedies to cure a febrile planet. They bring me round to the report's foreword – often as good a place to end as start – written jointly with the United Nations and the World Meteorological Organization. The emphasis in italics is mine.

> This Special Report confirms that climate change is already affecting people, ecosystems and livelihoods all around the world. It shows that limiting warming to 1.5°C *is possible within the laws of chemistry and physics* but would require *unprecedented transitions in all aspects of society* . . . Without increased and urgent mitigation ambition in the coming years, leading to a sharp decline in greenhouse gas emissions by 2030, global warming will surpass 1.5°C in the following decades, leading to irreversible loss of the most fragile ecosystems, and crisis after crisis for the most vulnerable people and societies.

Such is the human voice of science, exercised and prophetic. It can tell us what is happening and what is possible. But it cannot tell us what to do.

Reaching for a Lifeline

Against this dismal backdrop there is a glimmer of hope. Like we saw in the IPCC's report on land and climate change, SR1.5 zeroes resolutely in on 'sustainable development'. It pulls the science firmly

round and situates it in the concept that the Brundtland Report of
the World Commission on Environment and Development defined
in 1987, and summed up with its title, *Our Common Future*. In one
word, and one we shall come back to, we might say that the hope
lies in *community*. This shows in the four-part structure of SR1.5's
summary for policymakers. Part A looked at understanding global
warming. Part B at climate change projections, impacts and risks.
Part C at emission pathways and transitions for 1.5°C. And Part D
at strengthening the global response in the context of sustainable
development and efforts to eradicate poverty.

A one-page graphic summarises how each of the seventeen
Global Goals in the UN's 2030 Agenda link to energy supply,
energy demand and impacts on the land as the ground of human
living. The smaller print states that 2030 sustainable goals 'provide
an established framework for assessing the links between global
warming of 1.5°C or 2°C and development goals that include poverty eradication, reducing inequalities, and climate action'. These
call for capacity-building across government, civil society, business
and local communities including indigenous peoples, with international cooperation being a critical enabler for vulnerable parts
of the world.

If these are not put in place, or if programmes are poorly
designed or badly implemented, then with *high confidence* greenhouse gas emissions, gender and wider social inequality, and the
health of both people and ecosystems will suffer. But equally, if
they can be put in place, 'collective efforts at all levels, in ways
that reflect different circumstances and abilities' can strengthen the
global response to climate change, 'achieving sustainable development and eradicating poverty' – and that also with *high confidence.*

There we have it from the horse's mouth. It's down to us, or else
we're screwed. Let's now move on from climate science, and before
we round upon community, and the empowerment of peoples
from both bottom up and top down in synergy with the inner life,
let's start by looking at some things in our psychology that screw
us up.

SCEPTICS AND THE PSYCHOLOGY
OF DENIAL

The first day that I woke up to climate change, emotionally woke up I mean, was in 1985. I was no longer working out in the remote areas of Papua New Guinea. Instead, I was in Port Moresby, the capital city, having since done an MBA both to make myself more useful and to get a better insight into drivers of the problems of our times. My new role was as financial advisor to the South Pacific Appropriate Technology Foundation. Set up as an independent agency by the government after independence in 1975, inspired by Schumacher's principles in *Small is Beautiful*, its guiding star was 'integral human development' as the first directive principle of the new nation's written constitution: 'We declare our first goal to be for every person to be dynamically involved in the process of freeing himself or herself from every form of domination or oppression so that each man or woman will have the opportunity to develop as a whole person in relationship with others.'[1]

The real life backdrop was less idyllic. Young men had flooded to the cities hoping to find jobs, high status and a movie way of life – everything that advertising promised – which is to say, as one young man put it to me, 'all the fruits of the white man's garden'. When those proved far beyond their reach, the sociology of anomie, of alienation and degradation, often found expression through a drift into the city's *raskol* gangs. Even in the close-knit circle of our small Port Moresby Quaker meeting, we had experienced, gallingly, at first hand the death threats, robbery at knifepoint and gang rape.

It brought home to us that integral human development, an idea that had grown out of liberation theology and Catholic Church reforms of the 1960s, was more than just a nice turn of phrase. It was an urgent call to full humanisation. A reversal of all that had become dehumanised with the sudden loss of cultural anchoring that abrupt exposure to modernity had brought about. The technologies with which we needed to work were not just 'hard', albeit appropriate, technologies like micro hydro, solar crop dryers and hand-driven coffee husking machines, but also 'soft' technologies of how to live and work 'as a whole person in relationship with others'. That is to say, in community one with another.

Yet Another Bloody Thing

The director of the South Pacific Appropriate Technology Foundation was Andrew Kauleni, a Melanesian (as the ethnicity of that part of the world is known) from the Sepik River, a region rich in artistic and ceremonial life that borders onto Indonesian Papua. One day we were expecting him back at the office after a meeting of South Pacific leaders in Tuvalu. Formerly known as the Ellice Islands, this cluster of nine inhabited coral atolls makes up a tiny Polynesian nation of just 11,000 people. He walked very slowly into the office later that morning. A deliberation in his gait signalled that something was wrong and drew several of us round as he settled down behind his desk.

'If what we've just been told is true,' he announced, as if a judge raising the gavel, 'then low-lying islands like Tuvalu are going to disappear beneath the waves.'

For several years, the 'possibility' of climate change had been creeping up around the agenda of development agencies like a choking vine. It had been seeded by the World Meteorological Organization that had mounted the first World Climate Conference in 1979, resulting in a call on governments 'to foresee and prevent potential man-made changes in climate that might be adverse to the well-being of humanity'.[2] Most of us on the ground just didn't want to know. As development workers, we were about people, not ecology. Ecology was birds and bees. But AIDS had just arrived in

Papua New Guinea, and we were more worried about emergent stresses on the fledgling public health service. I can well remember feeling – 'Yet another bloody thing!' – as Andrew's words landed with a thump into my mind.

'If what we've just been told is true . . .'

Tuvaluan Poppycock?

But it wasn't true.

At least, not in Tuvalu's circumstance to date. In its review of climate change and the land, SRCCL devotes a case-study box to migration in the Pacific. Some Pacific islands are, indeed, experiencing displacement of their populations. On Yap, in the Federated States of Micronesia, extreme weather events over very low-lying land 'are affecting every aspect of atoll communities' existence'. However, notwithstanding the absolute rise in global sea levels, one study drawn on by SRCCL suggests that Tuvalu has actually experienced a seventy-four hectare net increase in land area. How so? Because, it answers, 'islands are dynamic'. Like we saw at Leurbost, but for tectonic reasons other than glacial equilibrium, the earth's crust at Tuvalu appears to be lifting at a rate that outpaces rising tides. The poster images of the island group 'as a laboratory for global climate change migration' may only be, as the IPCC quotes from recent studies, 'visualisations by non-locals', and even a case of 'wishful sinking'.

It's not hard to understand how this has happened. Back in the 1980s the threat of climate change was very new to much of the international community. The broad-brush science of sea level rise was only just starting to fall into place, and without sufficient data to distinguish regional or relative sea-level changes from the global or absolute level caused by warming. SRCCL concludes that 'climate stressors threaten the life-support systems of many atoll communities'. It warns that 'low adaptive capacity could eventually force these communities to migrate'. But the fact that in the early days somebody must have extrapolated beyond a reliable evidence base for Tuvalu had set a hare running. Forgivably in this case, the climate change denial lobby sounded the horn, but then set loose the dogs.

One master of the hounds was the late Clive James, a well-known author and humorist, who wrote a chapter for an anthology called *Climate Change: The Facts 2017*. This was compiled by the Institute of Public Affairs, a conservative Australian lobby group disguised as a think tank that describes itself as 'the voice of freedom', and built its pedigree in agitating for the deregulation of the tobacco industry.[3] The institute's accounts do not disclose the sources of its A$5 million revenue, but much is thought to come from the mining billionaire, Gina Rinehart.[4] In his contributed chapter, James described himself as 'a keen observer of popular culture and unsustainable fads ... who knows nothing about the mathematics involved in modelling non-linear systems'. He attributed the notion of 'climate justice' to Robert Mugabe's quest 'in which capitalism is replaced by something more altruistic'. Leveraging the case of Tuvalu he said, in the trademark snarky register of climate change denialism:

> I myself prefer to blame mankind's inherent capacity for raising opportunism to a principle: the enabling condition for fascism in all its varieties ... Later operators lack even the guilt. They just collect the money, like the Prime Minister of Tuvalu, who has probably guessed by now that the sea isn't going to rise by so much as an inch; but he still wants, for his supposedly threatened atoll, a share of the free cash, and especially because the question has changed. It used to be: how will we cope when the disaster comes? The question now is: how will we cope if it does not?

Tuvalu these days has slipped from poster case to whipping boy. Tiny though it is on the greater scale of things, it has become a wide open goal for climate change deniers. It has allowed a comfortably padded armchair flâneur like James to dismiss climate science as nothing more than 'fashionable nonsense', indeed, as 'poppycock'.[5] But against this backdrop, both absolute and relative sea levels continue to rise on many other Pacific atolls. Low-lying land providing food and space for settlements is also threatened by extreme weather events, storm surges and changing rainfall patterns that, by reducing hydrostatic pressure in freshwater aquifers, can cause the ingression of salt water into village wells.

The privileged watching from afar won't need to worry much. They've not written off the South Pacific, for not all its islands are low-lying. So it is that billionaires like Peter Thiel, linked to PayPal and Facebook, see New Zealand as 'the future'. Like other stable countries of a temperate climate and low population pressures – Canada, Scotland and Ireland amongst them – New Zealand has become for Silicon Valley's scions a 'favored refuge in the event of a cataclysm'.[6]

Climate Change Denialism

I have had many run-ins with those who might loosely and to varying degrees be described as climate change deniers. Most of these have been on social media or face to face at meetings and debating panels. Invariably, in my experience, they have been white, male and middle class, and I usually get the impression, unwilling to consider any restraint upon their lifestyles. This often comes with a narcissistic presumption of entitlement that, if challenged, hints at a brooding anger; a resentment that, I cannot help but ponder, might have more to do with early childhood issues than with any real debate about the science.

In one discussion on BBC Radio I'd gone along to the studio, having prepared for an in-depth exploration of global warming's implications for the way we live. Instead, and without warning, I was pitched against a climate change denier. He squandered all our time on flippant whataboutery that shed little light and only luke-warm heat. It might have been some entertainment for the audience, but did nothing to inform. I came away a little disappointed by my own bewildered response. Had I been forewarned, I could have been more nimble on my feet and used the space, with due respect, to expose denier tactics.

Another setting was an online debate in 2010 that *ECOS*, the journal of the British Association of Nature Conservationists, organised between me and a wildlife ecologist, Peter Taylor. Taylor's

2009 book, *Chill*, argued that far from living in a world that's heating up, 'the period 2002–07 marks a turning point, then glaciers will begin to grow and ice mass begin to accumulate again, thus levelling off the sea level rises'. He saw the cold winter of 2008–9 as heralding the coming ice age.[7] Being an ecologist, this made him a hero of climate change denialism, an avid convert from the other kirk, and for a time, *Chill* ranked as number one in Amazon's bestselling league for 'global warming'.

Invariably I have found myself asking of such figures, who have no credibly peer-reviewed publications in climate science: what makes them think that they know better than experts with a reputation worth not losing? I ask myself what drives such attitudes. Taylor concedes that the heavy impact of climate mitigation measures on nature and landscapes influenced his views. It's something that I've often seen in question time at public events, and also, with my late friend the ecologist David Bellamy. Folks who fiercely love the countryside can't bear to see wild places bulldozed through with tracks to put up wind and solar farms on a jangular (as I call it – jarringly angular) industrial scale.

Taylor praised our *ECOS* exchange. He said at the time: 'I know of no other consistent debate on this important issue.' Not having been in touch for years, I dropped him a line again while writing this chapter. I asked: given that his forecast 'chill' has not materialised, did he think that it was coming yet, for all that? His reply was characteristically avuncular. It left my question feeling almost mean-spirited. He made no reference back to his previous predictions. Instead, to my astonishment, he wrote of 'record warmth – just as we could expect', that the current warm period 'may have two or three centuries to run', and the next ice age is not just around the corner but 'three to four hundred years away'.[8]

It seemed that the denial had full astern gone retrograde. I scratched my head and gave a weary nod to all those hours spent on the great debate. Probably no point in getting back to say that the effects of current CO_2 concentrations will likely linger on for many thousands of years, and not just a few hundred. In my mind I wished him well. Every one of us is on a massive learning curve in this game. And I left it there.

Climate Change Contrarianism

Most public figures who confront or confound the science and resulting policy measures no longer say that global warming isn't happening. Instead, they adopt a 'contrarian' position, also known as a 'climate dismissive' position, taking issue instead with abstruse elements of the scientific data, with the extrapolated rate of heating, with the attribution of its causes or with the expected impact and anticipated costs – not least the 'socialist' taxation and regulatory implications – of actually doing something.

The Global Warming Policy Foundation is Britain's foremost 'climate sceptic' lobby group. It was set up by Lord (Nigel) Lawson of Blaby, Mrs Thatcher's former chancellor of the exchequer, and its website – literally a 'dark' web in its presenting colour scheme – boasts a formidable board of heavyweight political figures, contrarian scientists and erstwhile captains of commerce, the media and the civil service. To see power at work – elevated, concentrated and networked – go no further than to take a look online, and gape.[9]

Lawson refuses to disclose the sources of funding, conceding only that he relies on friends who 'tend to be richer than the average person and much more intelligent than the average person'.[10] An indication of the type of person this might be is that in December 2019 the Foundation appointed as its new chairman Terence Mordaunt, the co-owner of Bristol Port, a cargo-handling business that is critically positioned to benefit from any increase in trade with America. Mordaunt's estimated worth exceeds a third of a billion pounds, he is a major donor to the Conservative Party as a backer of Boris Johnson's leadership campaign, and gave £100,000 to the official Brexit movement that campaigned for Britain to leave the European Union in the 2016 referendum.[11]

I encountered the Foundation's director, Benny Peiser, in a panel discussion in 2013 at the HowTheLightGetsIn philosophy festival in Hay-on-Wye. He did not deny warming. Instead he argued that the future lies in an advanced economy and 'high and risky forms of technology' – geoengineering, I inferred – because that's what it means to be a modern society. Green policies, he insisted, are irrelevant and even amoral. They ignore the needs of millions of people in want of development, and are little more than another affectation of 'people who have the luxury to put a £20,000–£30,000 solar panel on their home'.[12] It sounded like a passionate appeal for social justice.

My riposte was that our sixteen solar panels, even in rainy Glasgow, produce just over 3,300 units a year. That's approximately the same as the average domestic electricity consumption of a UK household – and free from off the roof. The panels cost us £5,000, installed without any capital investment subsidy in 2013. Having set them up to use the power in combo with an air-to-air source heat pump, costing £2,000 installed, we've slashed the carbon impact of our gas and electricity. Seven years have now confirmed that we have reduced our domestic carbon footprint from 5.4 tons of CO_2 per annum to an average of 1.95 tons, a fall of 64 per cent. The capital investment paid itself off after just five-and-a-half years. In terms of Peiser's expression of concern about poverty, we live on a nice street in a hard-pressed area and surrounded by social deprivation. The entire terrace is south facing and all the roofs are similar, but only one other home has panels fitted. Our fuel bills now come in at £750 a year. That's about half of what they were, even without accounting for price increases since 2013. Subtract from that the solar 'feed-in tariff' payment that we receive of around £650 on each year's production, and it ends up that we pay £100 net a year for all our gas and electricity. *One hundred pounds a year.*[13] Even if you take off the subsidy element of the feed-in tariff (which no longer exists for new installations), it would still be well worthwhile given the patterns by which a household such as ours uses energy.

This illustrates that not only can green policies help to save the environment, they can also, in the right contexts, combat fuel poverty. Just ask in the social housing around our neighbourhood that the Scottish Government have insulated, and they'll know.

Peiser had no answer to my riposte. Neither did he seem to have a plan for countering the global rise in CO_2. As for the consequences of climate change, when asked by the *Sunday Times* about the implications of sea level rise, he reportedly replied: 'The predictions come in thick and fast, but we take them all with a pinch of salt. We look out of the window and it's very cold, it doesn't seem to be warming.'[14] His bottom-line positioning seemed to be for business as usual, where increasingly for large parts of the world, there may be no more usual.

One might imagine the appeal of this to the intelligence of Lord Lawson's 'richer than the average person' friends, who enabled the foundation to have a turnover of a third of a million pounds in 2018. Equally appealing to them would be the work of Andrew Montford, a chemist turned chartered accountant, also without a background in peer-reviewed climate change science.[15] In 2017 Montford was appointed deputy director to work alongside Peiser. The following year he wrote its policy briefing paper that promoted shale gas and defended the fracking industry from what he sees as green media spin.[16] My encounter with him came in 2010 when *The Scottish Review of Books* asked me to review his investigative work, *The Hockey Stick Illusion: Climategate and the Corruption of Science*.[17] Like Taylor's *Chill*, it was a book that quickly achieved cult status amongst climate change deniers. I concluded that at best it might help to keep already-overstretched scientists on their toes. At worst, it was a yapping terrier worrying the bull, one that cripples action, potentially costing lives and livelihoods.[18]

Montford runs a blog from which, under the pseudonym of 'Bishop Hill', he lampoons the high priests (as he sees them) of climate science and all such hooey as green taxes, subsidies, legislation and self-righteous preaching from the likes of, well, yours faithfully. His Grace, as his congregation deferentially refer to him, responded to my piece with two blogs that had me tossed into the dungeons of the Inquisition for heretical impertinence, an abomination unto the sensibilities of the Lord. A crusade was launched, a jihad ensued, and fusillades were fired from keyboards poised in every corner of his parish. In all, some 150 comments linger as remnant landmines on the good bishop's website.

'He is an enemy of the people and the state and is declared anathema,' said one. I took the humour as a badge of office. Even better, said another: 'Deploy heavy *ad hominem* artillery to characterize [him] as a coprophagic protocranial.' Verily, it's a sorry day when a literary reviewer has to go and look up even simple dictionary words. 'Adopt a lordly disdain and ignore him.' 'He and his eco-chums are in it for the money.' 'Another one of these weird Highlanders who seem to dominate Scotland.' 'Alastair, just keep tossing off your caber.' 'Yer Grace, show no quarter, none will be given.' 'He deserves a kicking.'[19]

I came out of such a Punch and Judy show well able to brush off the laugh. But it was all right for me. I make use of climate science coming from the early background of just a general earth sciences degree. I pitch no claim to be a climate scientist. Others, at the heart of science, suffer for their work. No quarter is the order of their day.

Scepticism Weaponised – Climategate

Montford introduced his book as bouncing off (and leaning heavily on) the blog site of a Canadian mining industry economist, Steve McIntyre, who has made an occupation of attacking the work of Michael Mann, professor of atmospheric science at Penn State University. It was Mann who derived the famous 'hockey stick' curve that historically reconstructs world temperatures for the past millennium, showing how they turn upwards very sharply around the start of the twentieth century. There is no serious scientific dispute around the well-replicated core assertions of Mann's findings, but because the hockey stick so graphically conveys the reality of climate change, it became the bogey for attack from across a spectrum of contrarians and outright deniers. These claim that the Medieval Warm Period, from around 900 CE to 1300 CE, proves that the curve should really be a U-shape, and that this demonstrates that the notion of anthropogenic global warming is a hoax – or as Montford said, 'the corruption of science'.

Put generally, the fundamental error in such an argument is that the Medieval Warm Period was mainly regional, not global, and neither was it as pronounced as once thought.[20] As a detailed analysis under the heading 'The Montford Delusion' on *RealClimate* concluded: 'The only corruption of science in the "hockey stick" is in the minds of McIntyre and Montford. They were looking for corruption, and they found it. Someone looking for actual science would have found it as well.'[21]

Mann spent the next few years in and out of the courts defending well-organised attacks against his work, his funding and his university tenure. Perhaps even more caustic than what he underwent, and forming the final chapter of Montford's book, was 'Climategate'. This anonymous attack was on staff of the Climatic Research Unit at the University of East Anglia, a team that were working in similar fields to Mann. Thousands of their emails on the university's server were hacked, organised and published on the internet to make out that the scientists had tried to massage the Medieval Warm Period out of existence. Their release was timed for just before, and seemingly to sabotage, the meeting of world leaders to decide on climate change action at COP 15 in Copenhagen in 2009.

Professor Phil Jones, who led the unit, and his colleagues were subsequently put through multiple governmental, professional body and university investigations, all of which exonerated them.[22] Sometimes the wording in their emails had been colloquial, such as referring to a statistical method of data handling as a 'trick'. Such language is a norm of shorthand communications between co-workers who'd never think that such a word might be intercepted, and then played out to infer trickery.[23] For several years afterwards, Jones endured a stream of frightening and invariably anonymous threats towards his person, his wife and his children. It leaves one to wonder just what kind of people one is faced with, and how their motivations are both driven and resourced.

A decade on, and Climategate might seem like water under the bridge. The brutal rate at which temperature records lie in ashes around the world leaves few of even the most hard-bitten deniers still intent on trying to kink the hockey stick. However, climate scientists across the board have been left scarred by the iconic incident. There has been a tightening, a palpable reluctance amongst many, to engage in public contexts that might descend into the sport of bear-baiting. A study by Rosemary Randall and Paul Hoggett into the emotional impacts of being a climate researcher describes respondents testifying to having become afraid of putting their heads above the parapet. One remarked: 'There are people out there who watch everything that I say . . . so I feel very threatened and intimidated, and you see it's changing my behaviour.'[24] Spencer Weart, a science historian with the American Institute of Physics, put it like this to the *Washington Post*: 'We've never before seen a set of people accuse an entire community of scientists of deliberate deception and other professional malfeasance. Even the tobacco companies never tried to slander legitimate cancer researchers.'[25]

Part of the pity of this is that respectful and informed scepticism should, and does, play a vital role in science. Fraud is less likely in today's world where most scientists work in large teams with intense and competitive formal and informal peer review. Even then, however, error can sometimes dribble its way past reputably peer-reviewed goalposts. A pointed example was in 2018, when a climate-sceptical English mathematician, Nicholas Lewis, who has both written for the Global Warming Policy Foundation and been an IPCC expert reviewer, picked up on a statistical mistake in a paper about exacerbated ocean warming that had been published by *Nature*. Within two months, the journal flagged up a potential problem and subsequently retracted the study.[26] That's exactly how science should work, though as Lewis fairly pointed out in a blog comment, 'it is also very important that the media outlets that unquestioningly trumpeted the paper's findings now correct the record too.'[27]

We might note that Lewis was coming from within his field of competence. He thereby rendered science a service. But that is an exception rather than the norm. According to one study undertaken by climate scientists and journalists, whereas 97 per cent of published scientific papers accord with the general consensus on anthropogenic warming, the remaining 3 per cent that take a contrarian position tend to be blighted by shared flaws such as cherry-picking evidence, lifting material out of context, flawed methodology, dodgy fitting of data to curves and 'disregarding known physics'.[28]

Where scepticism slips over into hard-line denial, there is no such healthy calling to account of what are probably the alter-egos of some otherwise quite ordinary if sardonic people. Instead, we run into the sneering vim and vigour, sometimes the mocking sadism, of anonymous bloggers, columnists and letter writers, whose cultivated buffoonery can carry with it quite a following. As Lucy

Mangan asked in the *Guardian*, writing on the tenth anniversary of Climategate: 'Is it pure arrogance that makes laypeople think they know better than scientists who have spent their lives painstakingly researching an issue? Or a desperate insecurity that makes them unable to stand the respect accorded to experts?'[29]

Well-crafted denialism exploits such insecurity. Masquerading as just the honest doubter, it hooks into the Microsoft security vulnerabilities that we all have in our psyches. It appropriates the role of 'sceptic' as an honourable badge of office, but in ways no different, no less ego building and sometimes no less lucrative, than the quack practitioner who subtly sows 'alternative facts' to undermine a cancer patient's confidence in evidence-based medicine. Such denialism simplifies reality and twists it, not to science, but to a pleasing storyline. It purveys a reassuring narrative that's well within folks' sense of being 'in control', a confidence that can be passed around in pub or club as if they understand it all. It keeps at bay what might be fears, guilt and a sense of shame, not to mention all that lurks behind a need for CO_2-belching markers of identity such as wait out in the car park round the back.

Such are, the truth is hard to tell, the lotus-eating holidaymakers who hire a boat and put to sea, not knowing how to sail. At best, their ports of call – 'our' ports of call, because this can be about us all to varying degrees – will be along a sharply rising learning curve. At worst, well, put it this way, there are Sirens on the rocks who play beguiling music. Not for nothing was the anti-Greta, Naomi Seibt, hired by the American conservative libertarian think tank, The Heartland Institute, in January 2020. Her role – to project to youth her message that climate consciousness is 'a despicably anti-human ideology'.[30]

Mass Media False Balance

On these matters of learning and light entertainment, not surprisingly, it can be hard for media editors and journalists to navigate contested discourses. Such terms as denier, contrarian and sceptic are all used in ways that accountants would call 'fungible', meaning that they glide into one another; and all lay claim to being the real science. Most people in the media have backgrounds in the arts, not the sciences. Climate change dismissives exploit this in the name of 'balance', appealing for 'equivalence' in public debates, but disregard the very real asymmetries that a little know-how with an axe to grind can represent.

The BBC provides a good focus, not because it is especially culpable – far from it relative to the privately owned press – but because it comes under such high scrutiny and requirements for accountability as a public service broadcaster. In 2014 the Corporation received sharp criticism for 'false balance' in a report called *Communicating Climate Science*, produced by the Science and Technology Select Committee of the House of Commons. This found that the BBC had exercised poor discrimination in assessing the level of expertise of contributors to its programmes, and was 'disappointed to find it lacked a clear understanding' of the needs of audiences in making sense of climate science. The committee heard evidence that climate change dismissives often grab attention by painting themselves as David in a fight with Goliath. The *Daily Telegraph* and the *Daily Mail* were criticised too for their failure to distinguish between fact and opinion, with opinion pieces 'frequently based on factual inaccuracies which go unchallenged'. Part of the problem lies in what constitutes 'news'. This, because:

> Media reporting thrives on the new or controversial. We heard that it was difficult to justify news time maintaining coverage of climate science where basic facts are established and the central story remains the same. Reporting on climate therefore rarely spends any time reflecting on the large areas of scientific agreement and easily becomes, instead, a political discussion on disputes over minutiae of the science or the policy response to possible impacts of climate.[31]

In April 2018 Ofcom, the UK's broadcasting regulator, found that the BBC had breached broadcasting guidelines by 'not sufficiently challenging' Lord Lawson's assertion on Radio 4's *Today* programme that 'official figures' showed that world temperatures had actually 'slightly declined'. In September, Fran Unsworth, the BBC's director of news, issued new guidelines to its journalists. These conceded that the Corporation had been getting its coverage 'wrong too often', and that, 'Manmade climate change exists: If the science proves it we should report it.' To avoid false equivalence, 'you do not need a "denier" to balance the debate'.[32] In other words, a round-earth programme doesn't need a flat-earth voice.

This new emphasis did appear to have taken root early in 2020 when – following a swathe of extreme events ranging from the Californian and Australian bush fires to unprecedented flooding in Jakarta – the BBC announced a year of special programming. The theme, *Our Planet Matters*, would lead up to COP 26, originally anticipated at the end of the year (before COVID-19 struck), and in educative ways that seemed to have taken on board the concerns raised by the select committee.[33]

Psychology of Cognitive Dissonance

What are the deeper drivers of denial in the face of science? Honest doubt and healthy questioning of the dominant paradigm is one thing, but I keep thinking of a discomfiting Persian proverb: 'Behind every rich man is a devil. Behind every poor man are two.'[34]

The rich have the devil we all know. But the poor have both the devil they might know and the one that might emerge, if given half a chance, were they ever to become rich. We must be careful with that proverb. It can be misused to deny structural iniquities and to set up a false equivalence of power between the rich and the poor, between those who take the rent and those who pay it, and between those with options in life and those bereft of any. With that caveat, the great Swiss psychotherapist Carl Jung said that we all have a 'shadow' side. We all have parts of who we are that we hide, even from ourselves – especially from ourselves. The work of psychological growth, he said, is mainly at the coal face of the shadow, to turn coal dust into the gold dust of awareness. If we don't rise to the call of this great work of life, the shadow denied will be the shadow that trips us up. The devil – and what would we do without 'his' metaphoric help to see ourselves? – is 'diabolic', precisely because 'he' throws a diabolo, a flying disk, across our path from deep inside.

These things happen at those levels of our psyches – our totality of being – that are both individual and collective, realms that Jung called the 'collective unconscious'. We may not be aware of what it is we move inside of – the personal mental fields, the social fields and even deeper fields of being that spiritual psychologies speak of as being 'transpersonal'. The spiritual – that cross-personal and superconscious level of profound interconnection – may or may not be 'for real'. That debate need not concern us for now. What does concern us, is to hold in mind the possibility that the fullness of reality may be bigger than the small self, the ego, is generally pleased to entertain. And that this may affect our perceptions and interpretations of reality.

To remain, for now, at the interface of personal and social psychology, we might note some parallels between both climate change denialism (and alarmism) and an attraction to conspiracy theories. Conspiracy theories, being based on speculation built up from slender evidence, construct a worldview over which the holder's presumption of superior knowledge gives a chimera of narratorial control, and therefore, of relative agency and safety. Most of us only live the way we do because of science and technology. But if one's inclination or aptitude is not to understand these, the modern world can feel a very disempowering place, especially given the military and corporate abuses and alienation of such gifts. To accept not knowing, to appreciate the value of expertise, and perhaps to get more educated, requires both opportunity and humility. Humility is the prerequisite of knowledge. But that's not easy for the ego.

There is also a whole realm of denial theory, much of it built up in the mid twentieth century from research spurred by attempts to understand wartime atrocities. As Stanley Cohen puts it in a classic study, denial theory asks questions like: 'How could they do such atrocious things, yet think of themselves as good and decent people?'[35] Similarly so with climate change. Studies of 'deniers' suggest that many know that climate change is real, but actively avoid facing it. Some just lack the mental bandwidth to take it on. Others are confused by the debate, as when the media sets up a false equivalence of pitching a 'denier' against a 'believer', as if climate science raises questions not of fact but of faith. At the end of the day, a great many of us just don't want to hear any challenge to what might be the high carbon footprints that sustain our affluence and confidence of social standing in the world.

The psychologist Robert Tollemache conducted a series of interviews with middle-class Londoners. These asked about their resistance to facing climate change and cutting their carbon footprints. He received responses such as:[36]

> I don't like negativity. I don't like people moaning and groaning. I haven't got much patience, actually ... I don't like doom and gloom predictions, although I may think, *arggh*, there is probably some truth in what they are saying.

> We fly too much; erm, that is, I don't know what that is, that is being spoiled, having the money to do it and doing it.

> I don't feel guilty, no I don't feel guilty. I mean I, I, I sort of, I, I, put up a kind of, I put up a wall.

My friend the campaigner and researcher George Marshall made a landmark study of climate change denial called *Don't Even Think About It*. He concluded that, just as we are wired to take action to protect our welfare, so we are also wired to ignore threats that can seem too big to deal with. These disrupt our categorisations of the world. They become metaphysical issues, unsettling the very structure and stability of what gives constancy to reality. A parallel from zoology is the 'displacement activity' of an animal that, when under serious threat, may continue grooming itself as if nothing is happening: a kind of 'whistling in the dark' to head off things that go bump in the night.

Marshall sees denial as contributing to a socially constructed silence. So-called 'alternative facts' about climate science 'pollute the discourse' in ways that protect us from anxiety. We're therefore not so much fooled by the quack practitioners of climate science who belittle the threat, as much as 'we actively conspire with each other, and mobilize our own biases to keep it perpetually in the background.'[37]

Such thinking chimes in with cognitive dissonance theory, one of the most important bodies of insight that emerged from the mid twentieth century's flowering of social psychology. In a classic 1957 study of a flying-saucer cult, Leon Festinger and his team observed that people look for 'cognitive consistency' in their lives.[38] Even after the UFOs had failed to arrive and carry the sect's adherents off into some kind of techno-rapture, they carried on believing in what they had already committed themselves to. Some went as far as strengthening their belief, rationalising that the aliens, by not destroying the world in an apocalyptic flood, had only been testing their faith.

A more everyday example of such cognitive dissonance reduction – the need to reduce the gap between our chosen beliefs and factual realities – is what marketing grandly calls, post decisional cognitive dissonance reduction. Once we've decided on which new car to buy and signed the deal, we big-up its advantages and play down the disadvantages over rival models. We have a yearning for consistency between our beliefs, values, attitudes, intentions, behaviours and perceptions of reality. The dissonance or disharmony that arises when these are exposed as being out of alignment comprises the psychological discomfort of a clash of models of reality. This disturbance triggers primal feelings of insecurity that cognitive dissonance reduction strategies attempt to head off. If applied to climate change, such coping strategies include:

Reduced ascription of responsibility – 'Too big for me to make a difference.'

Denial of consequences – 'CO_2 is plant food, and I like a bit of warmth.'

Altering moral norms – 'Not my responsibility, it's the 1 per cent that's the problem.'

Shifting perceptions of reality – 'They'll find a way to fix it in the future.'

Projection of one's own shadow – 'These scientists, only in it for the money.'

Projection onto scapegoats – 'Too many Indians and too much coal in China.'

Dissociation from reality – 'Cold winter shock forecast in the Daily Express.'

Displacement activity – 'But we go to the recycling. We do our bit.'

Whataboutery – 'What about sunspots and the Medieval Warm Period?'

Telescopic philanthropy – 'I'd rather see the money spent on . . . .'

Apathy – 'It'll see me out.'

However, by dealing with it by not dealing with it in these ways, we succeed only in compounding the inner conflict between the kind of people that we like to think we are and have others believe in, and the secret self we really are. Notice in Tollemache's quotations above the contrast between the dislike of moaning and groaning and the *arggh*; between the spoiling of oneself and spoiling the world. And then there's the guilt that isn't there – except behind the wall. The price of such denial, such deficit of courage, is that we lose our grounding in both truth and how the world is. We sacrifice our authenticity as self-authored, integrated, individuated or self-realised human beings and our capacity to respond in truth to the way it really is, including what is happening to the natural environment that sustains all life. Jung saw the danger of such dissonance with a prescient clarity. He said: 'People who know nothing about nature are of course neurotic, for they are not adapted to reality.'[39]

The bottom line, is that climate change is not like familiar threats. It is not a simple or a single issue, like the ozone hole or acid rain, with relatively painless solutions in advanced technology and international law. Let me say it again: we've only built a world of nearly 8 billion people living the way many of us do because of the brittle hyper-efficiency of a just-in-time economy, powered by energy-dense fossil fuels. That's what makes cheap oil the life blood of the economy of globalisation. Climate change is not just symptomatic, an itch caused by an irritant. Climate change is systemic. Its drivers run through nearly every aspect of our lives. It highlights our neurotic relationship with reality – our being out of kilter with nature, self and one another – and it may leave few things ever quite the same again.

Beyond denialism, 'there's a killer on the road'. But this realisation, this recognition of reality, could be the silver lining. As Jim Morrison's prophetic anthem had it, 'the world on you depends / our life will never end.'[40] The very thing we might refuse to face might be our 'basic call to consciousness', as the Iroquois Six Nations told it to the United Nations. The means by which we become the riders on the storm.[41]

I think of that French wildfire mentioned earlier. How Vérène and I had driven through miles and miles of vineyards and mused on how much land the world gives over to alcohol production, or tea and coffee, or animal feedstock: things that if push ever came to shove are slack in the agricultural system and mostly not essential for survival. The way Joëlle, my mother-in-law, said that after grieving for their local woodlands, the thing that stayed most with her was the way that she and her husband Hubert had to think, with only minutes to get out, what things to take in that one suitcase they were told to pack.

What might we pack in ours? What valuables and values are worth picking up and taking forward? And how might these help, and with what practical implications for action and leadership in our social movements, as we heed the basic call of climate change?

REBELLION AND LEADERSHIP IN CLIMATE MOVEMENTS

Let me repeat a line from earlier: *Mostly, we only know what we think we know about climate science because of the climate science.* This might give cause for reflection both to climate deniers and, at the opposite though much more thinly tapered end of the spectrum, to alarmists, where climate activism can also edge out of step with the science. When this happens, activism can set itself up as a straw man that climate change dismissives can readily knock down, 'because they exaggerate'.

Here I will continue from the scientific basis discussed at the start of Chapter 2: namely, one that privileges peer-reviewed, expert panel-appraised, consensus-settled science for all its limitations. In shorthand, that means IPCC science, and notwithstanding attempts by both deniers and alarmists to knock it, I hold the view that the IPCC is an incredible organisation. There is just nothing else like it in the world. The very process of working out their consensus pushes their science and their conclusions to a level of integration that is beyond merely aggregating their contributions. However, what the science says and how different groups use it is another matter. Having looked at climate change dismissives of varying shades, I want to look now at how the science plays out in activist climate change movements and especially in what has been called the 'obstructive programme' of protest. After that, in later chapters, we will touch implicitly on aspects of what Gandhi called the 'constructive programme' – that of building alternatives to systems based on domination and exploitation even while those systems continue to operate. I say, 'implicitly', because the options are so vast that I would rather show by example, especially from personal experience, and with nods to the UN's 2030 sustainable development goals, than by making out that I have a 'plan' or writing lists that kill the spirit.

In what follows, I will focus on Greta Thunberg and, especially, Extinction Rebellion, as the largest and most visible faces of climate change mass movement at the start of the third decade of the twenty-first century. However, many of the issues I will raise are generic to social movements, and so I would invite reflection on them more by way of patterns and examples. Individuals and organisations will come and go in the splash and dash of ephemeral life, but the groundswell of deep human learning takes place on a longer oceanic wave.

Greta Thunberg's Prophetic Activism

In August 2018, during Sweden's hottest ever summer and with a general election looming, the child activist Greta Thunberg threw herself into the School Strike for the Climate by protesting outside the Swedish parliament. As a young adult now, I will speak of her as 'Thunberg' and not presume the 'Greta' informality of a child. By March the following year, an estimated 1.6 million girls and boys from 125 countries had joined this latter-day Children's Crusade. As a plethora of weather records continued to be broken around the world, Thunberg was elevated to the world stage where she delivered a series of apocalyptic dressing downs.

At the World Economic Forum in Davos: 'I don't want your hope, I don't want you to be hopeful. I want you to panic. I want you to feel the fear I feel every day. And then I want you to act, I want you to act as if you would in a crisis. I want you to act as if the house was on fire. Because it is.'[1]

At the United Nations in New York: 'People are suffering. People are dying. Entire ecosystems are collapsing. We are in the beginning of a mass extinction. And all you can talk about is money and fairytales of eternal economic growth. How dare you!'[2]

To her admirers, here was the 'kick-ass' prophet in full declamation whose fire was righteous indignation. To her critics, here was the 'annoying' projection out onto the adult world of a 'good child, bad parent' petulance. That combination of both approbation and castigation, together with her steely determination armed with the science of SR1.5, rendered her a global news phenomenon. Properly speaking, a prophet is not a teller of the future, and universal popularity has never been in the job spec's remunerative package. A prophet's job description is to pay heed to their inner calling, to read the outer signs of the times and to speak to the conditions found upon the land to call the people and their leaders back to what gives life.

At the US Congress she tabled a copy of SR1.5 to be duly entered into the congressional record. Mostly, though not always, she stuck carefully to the consensus science, telling a top European Union committee that they were acting like spoiled children, and to 'unite behind the science, that is our demand'.[3] But how to square such science with the politics? As she summed it up in a tweet, to meet even just the commitments of the Paris Agreement: 'The politics needed . . . does not exist anywhere today. Not one single political party to my knowledge is even close to approaching the solutions required for the IPCC 1.5. But that is about to change. We, the people, have started new movements. Join in!'[4]

The remit of SR1.5 was only ever to advise upon the scientific options. Its expertise and calling lay not in what was socially and politically possible. In her Davos speech, Thunberg had the honesty to signal that she understood such limitations. She conceded that even to achieve the underwhelming Paris targets, governments would have to: '. . . include negative emissions techniques on a huge planetary scale that is yet to be invented, and that many scientists fear will never be ready in time and will anyway be impossible to deliver at the scale assumed'.

On the one hand, this determined young woman, as if an activation of the Joan of Arc archetype, quite properly called for action on the 1.5°C target. On the other hand, not anybody – not the scientists nor the politicians nor herself and the activist community included – had come up with a politically credible plan to achieve it. Was this demanding the impossible, or was it fair enough? After all, it is the prophet's job to tweak the whiskers of the king, not to assume the throne. The child calls out the emperor's new clothes, but to tailor the real thing – that's quite another proposition.

Extinction Rebellion's 'Demands' on Government

Extinction Rebellion – often abbreviated to XR, with the X stylised in the logo to suggest an hourglass with the time running out – was launched into the public arena on 31 October 2018. Thunberg spoke at this inaugural Declaration of Rebellion, saying: 'We can't save the world by playing by the rules. Because the rules need to change . . . It is now time for civil disobedience. It is time to rebel.'[5] It was in the same month as the IPCC had published SR1.5 and just a month earlier, António Guterres, the UN secretary general, had said: 'Climate change is the defining issue of our time . . . We face a direct existential threat.'[6]

Extinction Rebellion had been taking shape since April that year. Its co-founders were Gail Bradbrook, a Stroud-based coal miner's daughter with an interest in spirituality, Simon Bramwell, with an interest in regenerative cultures that can rebuild healthy communities, and Roger Hallam, a Manchester-born Welsh-based organic farmer and mature student of civil disobedience.[7] The fledgling organisation went beyond SR1.5's target of a global 45 per cent emissions cut by 2030, with net zero reached by 2050. By 2025, Britain will have used up its per capita fair share of the remaining global carbon budget to keep within 1.5°C,[8] therefore by 2025 it had to be for both the sustainability and the fairness between nations of a 'just transition'. To this end, the group laid three 'demands' before the government. We might sum these up as TAP – truth, action and politics:[9]

> Tell the Truth – government must tell the truth by declaring a climate and ecological emergency.
>
> Act Now – government must bring CO_2 emissions down to net zero by 2025.

Beyond Politics – government must create, and then be
led by, a Citizens' Assembly.

As the movement's manifesto handbook, *This Is Not a Drill*, would
put it a year later in a powerful expression of values: 'It becomes
not only our right but our sacred duty to rebel . . . We act in peace,
with ferocious love of these lands in our hearts. We act on behalf of
life.'[10] Ten guiding principles are laid down on the website, all driven
by the vision of 'a world that is fit for generations to come'.[11] They
include growing a 'regenerative culture' that is healthy, resilient
and adaptable; welcoming 'everyone and every part of everyone';
avoiding 'blaming and shaming'; and breaking down hierarchies by
building on nonviolent strategy and tactics, decentralisation and
equitable participation. 'Anyone who follows these core principles
and values can take action in the name of Extinction Rebellion.'
But this would give wide licence.

Come October 2019, more than 1,400 people were arrested in
London alone as roads and bridges were blocked.[12] Some groups
held that protests should be guided by 'smart targeting'. One
action at a London arms fair saw twenty-one activists arrested. In
Germany and Finland, kayakers blocked the path of cruise ships,
thereby landing the inconvenience on the most profligate polluters
in the leisure industry. However, other groups partook of the 'any
publicity is good publicity' school of public relations. Disruption
to London's underground railway system caused divisions within
the movement and disrupted public support. People asked: 'Aren't
we meant to be encouraging public transport?'

An anonymously funded report for a conservative think tank
by a former head of counter-terrorism in the London Metropolitan
Police suggested that Extinction Rebellion 'is an extremist organisation whose methods need to be confronted and challenged
rather than supported and condoned'.[13] But neither was it just
reactionary influencers that thought they saw a problem. Senior
figures within the movement itself conceded that lessons needed
to be learned: these, as one anonymously told the *Observer*, 'most
especially within our own internal decision making'.[14] Meanwhile,
Extinction Rebellion as a whole had rejected a proposal to ground

flights at Heathrow airport by flying drones within the perimeter exclusion zone. That notwithstanding, under the auspices of a splinter group called Heathrow Pause, Roger Hallam ended up serving a six-week jail sentence for symbolically attempting to do just that. As the movement grew, so the questions became more pressing as to how the anarchy of a would-be non-hierarchical structure, one that permitted a high level of autonomy in its name, might be, could be, or should or should not be organised.

Tell the Truth

Extinction Rebellion's driving principle is that if only the truth about climate change were to be known to the public, if only governments told the truth or if it was otherwise outed, people would embrace the cause with a whatever-it-takes enthusiasm, rise up and overthrow the existing corrupt political and capitalist economic system that feeds denial and blocks radical action. Implicit to this theory is that we are not individually responsible for the way 'the system' is. We have been victims of its circumstances, and therefore, a revolutionary and not an evolutionary transformation of social, economic and political structures is required. In this, the third millennium, a new 'millennium' in the sociological transformative sense of that word, a new world order must be born.

By what means? The methodology that Extinction Rebellion adopted was built on the observation that nonviolence has developed a strong track record of success. To *react* to oppression using the oppressor's tools merely gives the oppressor home advantage. You have to get better at doing violence than those who already specialise in the business. But if instead you *respond* to oppression, using nonviolent forms of confrontation, you can perhaps flummox the logic of violence and break its vicious circle. Political events around the turn of the millennium had already put nonviolence firmly on the table. A whole new generation had been born into the civilian uprisings that led to the fall of Ferdinand Marcos in the Philippines in 1986, the collapse of the Berlin Wall in 1989, Ireland's Good Friday Agreement of 1999, Burma's Buddhist-led Saffron Revolution of 2007–8, and the Arab Spring of 2010–11.

These movements ranged from the successful to the faltering and the stillborn, yet wrapped up within them was something of the power of smallness squaring up to the gargantuan. Many such approaches had roots in, or influences from, liberation theologies – Islamic, Buddhist, Christian and others – that owed much to feminist, post-colonial, landless, urban poor and ecological analyses of oppression and the processes of liberation from it into a full and perhaps God-given humanisation.[15]

Accordingly, the new millennium saw nonviolence assume a growing role in workshops and practice at activist gatherings and camps. The UK's presidency of the G8 summit in 2005 was kicked off with a government-sponsored international conference at Exeter University, Avoiding Dangerous Climate Change, that helped to set a climate tenor for international protest several months later at Gleneagles in Scotland. Figures like the neopagan activist Starhawk came and taught ecology and nonviolence to protestors, exploring ways of engaging with 'webs of power' in the 'global uprising', and how to organise in 'collaborative groups'.[16] Once the G8 was over, such advocacy continued to deepen, making the connections between economics, ecology, prejudice, poverty, war and climate change in ways that led to nonviolence workshops becoming regular features at events like Climate Camp, Occupy and the COP 15 protests in Copenhagen and beyond.

Come 2018, Extinction Rebellion's founding leadership – with their roots in anti-capitalist protest – had latched onto a heroic narrative by which they believed that nonviolent rebellion works. Being told the truth about climate change would motivate mass revolt. Nonviolent direct action would precipitate mass arrests. These would generate public awareness and sympathy for the cause. As more and more protestors made martyrs of themselves – especially the young and otherwise vulnerable – others would come out and replace them on the streets in what is called the 'backfire effect'. This renders state repression counterproductive. The more they come at you the more they show themselves up, and thus, in a martial arts-like manner, they undermine their own legitimacy. It had worked for other causes in the world, so why not for climate action? Pivotal to such strategy was a doctrine, written into the group's second guiding principle: of 'mobilising 3.5% of the population to achieve system change'.

Why such a specific figure? Because those who masterminded the 'theory of change' had been overwhelmingly influenced by a single book: Erica Chenoweth and Maria Stephan's scholarly study, *Why Civil Resistance Works*. In examining more than three hundred examples of popular uprisings against tyrannical governments of the previous century or so, the two American researchers were surprised to find that nonviolence had proven itself to be nearly twice as effective as violence. Moreover, when a peak protest event achieved the participation of 3.5 per cent of the population, success seems to have been inevitable. Beethoven would either roll over, or succumb to just a gentle shove. The inference for climate change would be that being told the truth would lead to obstructive civil disobedience, which would lead to escalation, which would lead to the tipping point of social change, which would inaugurate a new political system, which would initiate the necessary steps to terminate greenhouse gas emissions. Thereby the planet would be saved.

Act Now

Given the urgency of the situation, Extinction Rebellion's second demand is for governments to bring emissions down to net zero by 2025. However, in the organisation's handbook, *This Is Not a Drill*, the human rights journalist Hazel Healy asked: 'What if . . . we reduced carbon emissions to zero by 2025?' Not knowing her, I can but presume that the answers are hypothetical and illustrative, for her logic was as gruelling as it was impeccable:

> Energy would be stringently rationed, dedicated to survival and essential activities; we'd go to bed early and rise with the sun. Expect massive disruption in the way food is grown, processed and distributed . . . Globally, there would be much-reduced private car use, virtually no aviation, haulage or shipping – spelling a dramatic end to material globalization as we know it.[17]

The politics, for a window of time that was just over one term of the UK's Parliament, sounded equally gruelling:

> But how to enact change on this scale? To avoid a totalitarian, 'eco-fascist' dystopia, a zero-carbon plan delivered at this rate would need to be contingent on total buy-in, perhaps triggered by a sooner-than-expected suite of apocalyptic impacts such as the collapse of pollinating insects, severe typhoons and saltwater flooding.

I have put it to friends involved in Extinction Rebellion planning: 'Might this not be possible, only under an authoritarian government?'[18]

The answers followed edgily. One said: 'Maybe we need a little bit of that.'

I replied, reluctantly to such a committed person, 'Then I hope that you can do it very well. Because there are others out there in the world who have very different agendas. They'd be just delighted to have that bandwagon legitimised.'

Beyond Politics

Extinction Rebellion's third demand states: 'Government must create and be led by the decisions of a Citizens' Assembly on climate and ecological justice.' This reflects much wider social concerns that democracy, as we have known it, can leave many feeling alienated from a deficit of agency. Increasing political specialisation, managerialism and centralisation, the self-seeking arrogance of what can be an out-of-touch 'political class', and the unfair lobbying impact of plutocrats – *Plutus* was the Greek god of wealth – and corporations, who fund the perversion of democracy, have all conspired to undermine trust in the truth and integrity of public affairs.

If representative democracy is felt to be failing us, and if participative democracy is felt to be the best remedy for the context, then what of more hands-on processes? Once, when visiting Connecticut in the USA, I was taken to experience a 'town meeting' that was held in the tradition of New England local democracy.[19]

The community that evening had to make a budgetary choice between a new fire engine or repairs to their school. It was moving to hear the fire fighters, the teachers and the parents all make their pitches in what was a frank, painful, but mutually respectful, and therefore, a caring, process of discernment. You got the sense that whatever lost out with good grace this time round would have it made up the next time round. Decisions at such meetings are usually determined either by a show of hands, or by a headcount for close calls. It brings it all out into the open.

These processes can work well in small communities. They can work too in wider contexts where most participants have sufficient prior immersion and personal investment in the issues to be up to speed and to remain on board. An early example, of which I was a co-organiser, was the People & Parliament process when the new Scottish Parliament was being set up in the late 1990s. Here we engaged some 500 groups across the country in a national discernment process about identity, vision and desired political process through to 2020.[20] Another was the Citizens' Assembly of Scotland of 2019–20 that tackled the primary question: 'What kind of a country are we seeking to build?'[21] A similar process was Democracy Matters that explored peoples' wishes for governance in England.[22] An official Extinction Rebellion briefing on citizens' assemblies cites Ireland as a successful case study. Here the process assisted public discernment around the reform of abortion and same-sex marriage laws.[23]

Just to hold a citizens' assembly or similar participative process does not guarantee that it would come up with an outcome that advocates of the process might desire, or for that matter, which would be acceptable to non-participants. An example of a well-run process that is often cited in the literature was the 2007 Ontario citizens' assembly on electoral reform.[24] However, when the proposal from its 103 members was put to the electorate in a referendum, it was defeated by a 63 per cent majority, which left the status quo in place. The scholarly literature suggests that while participative processes can help to bridge the democratic deficits left by representational democracy, they are not a replacement for it. This is important to understand, otherwise participants can be given a sense of false agency. On that point, I once served on a truth and reconciliation commission. It resulted in some good, targeted work that had positive legislative impact in the country concerned; but in its wider framing, I sometimes felt uneasy about listening to testimonies on issues over which we had little locus or capacity to intervene. While a sympathetic ear can be precious in its own right, it felt at times like encouraging folks to put their foot on the accelerator, but all the wheels could do was spin round in the sand and kick up dust. The linkages or lack thereof between three things must be explicit: the deliberation over decisions, the agency to implement them and the carrying of responsibility for the consequences.

Citizens' assemblies on climate change therefore need to be very clear about what interface they do or do not have with the legislative processes of representative democracy. If recommendations or decisions are being made on an assumption that the assembly reflects the wider population, it should be remembered that even a representatively selected assembly of 100 people will have a statistical error margin of about 10 per cent. In expectation management, it is important that organisers are aware not just of the strengths of participative processes but also of the weaknesses and criticisms. These include selection bias towards people who actively wish to have a public voice, the self-filtering of participants according to whether they have time to engage, paid or unpaid, with a prolonged in-depth process, and variations in capacity to understand the complex issues that climate change brings up. One Swiss study concluded (in the particular context that was being looked at) that citizens' assemblies can be more inclusive in some ways, such as the socio-economic spread of their members, but in other ways ballots were more inclusive, such as age and gender spread.[25]

The biggest issue is that an effective citizens' assembly on climate change would have to have to be explicit in its remit where sovereignty resided. In my observation, partly as an invited speaker at Extinction Rebellion events in Scotland,[26] many people had expectations that the citizens' assembly they hoped to see would have legislative power over climate action policy. However, Extinction Rebellion's guide to citizens' assemblies that came out in June 2019 walked back this notion. It took what might be the more realistic position that a climate assembly's recommendations 'could' be rendered binding on Parliament if they were backed by at least 80 per cent of the assembly's members. A lesser number 'could' require Parliament to debate the matter.[27]

Such a process could certainly augment representational politics. It would leave assembly members better informed and probably more engaged as citizens. Its recommendations might add legitimacy and direction to subsequent democratic process. All these are helpful, but as mainly a consultative measure they hardly go beyond politics. Neither, for that matter, did the Climate Assembly UK set up by the British Parliament with a series of meetings over four weekends in the first quarter of 2020.[28] The process was clearly stimulating for the 110 selected participants who enjoyed inputs from a raft of prominent speakers, including Sir David Attenborough. However, it was criticised for framing the discussions and advice being sought within the 2050 net zero targets of the Paris Agreement. As such, it fell well short of the hopes that Extinction Rebellion had held in demanding it.

Far Beyond Politics

To go 'beyond politics', one has to turn elsewhere within the broad church that Extinction Rebellion can be. Before embarking down that route, let me be clear that I mean 'broad church' in a flattering and not a cynical way. Extinction Rebellion's supporters include, for example, the environmental lawyer, Farhana Yamin. As a representative of small island states, it was she who was widely credited with persuading world governments to cement the 1.5°C target into the Paris Agreement at COP 21. In an April 2019 protest, she super-glued herself to Shell's headquarters because she considers that Extinction Rebellion gives better expression to climate concern than the supposedly competent authorities.[29] Another prominent supporter is Sir David King, a former UK government chief scientist and special representative on climate change, who issued a court defence statement, explicitly based on the IPCC's science, on behalf of five Extinction Rebellion members in January 2020.[30]

But to go beyond politics, in August 2019, the most prominent co-founder, Roger Hallam, released a manifesto called *Common Sense for the 21st Century*.[31] While not claiming to be an Extinction Rebellion publication, the conflation was obvious and the document was actively promoted on the organisation's website. The *Guardian* columnist George Monbiot tweeted: 'If you want to understand XR, or how to change the world, read it.'[32] Hallam insisted on a shift from 'reformism to revolution'. He specified this meant 'radical collective action against the political regime which is planning our collective suicide'. Making his point in the form of a dialogue, he wrote:

Question: 'What do we want?'
Answer: 'We are going to bring down the government.'

Hallam went on to note that there is 'a deep psychological attraction to going into the unknown', and claimed the very fact that revolution sounds incredible makes it credible. This 'releases enormous political energy and imagination' and is in turn the motive force that 'has powered the rapid explosion of support for Extinction Rebellion'. Once the government is hobbled, 'there needs to be a post-revolutionary plan otherwise chaos will ensue'. Here is where the national citizens' assembly comes in. This will 'take over the sovereign role from a corrupted parliamentary system' and 'deliberate on the central question of our contemporary national life – how do we avoid extinction?'

A 'sovereign role' means one that has governing authority. As such, 'the National Citizens' Assembly will become *the new governing body of the UK* and will deal with the climate crisis' (my italics). The eighty-page document concludes: 'Friends, there are no easy options anymore. There is only one way that leads to true self-respect – and that is Rebellion. Lets get to it.'

The frustrations that Hallam expresses with elective democracy are not new in political theory, they go back in the Western world to at least the Greek philosophers.[33] But to attempt to bring down a democratically elected government *before an electorate is ready for change through the ballot box where that is available to all*, is another

Riders on the Storm

form of hubris: that of getting ahead of the people. Far from demonstrating 'leadership', it is likely to sow the seeds of counter-revolution, probably ending in violence. We might think that 'they' who don't agree with us on climate change are brainwashed by the plutocratic control of the media and corrupted politics. Unfortunately, they might think from their vantage points that we're the brainwashed ones. We therefore have to work on change with one another at much deeper and personal levels, starting with conversations across the garden fence.

Saying this does not preclude protest and nonviolent direct action to confront the state, or other monoliths of power. Along with thousands of others I have many times protested at the Faslane nuclear submarine port twenty miles from Glasgow. On one occasion with some 300 fellows allegedly in 'breach of the peace', I was arrested, spent fourteen hours in a police cell, and was later put on trial for participating in a blockade of the main gates of the base.[34] But such protests as organised by Scottish CND and Trident Ploughshares are not attempts to bring down the structures of governance. On the contrary, they seek to assert the primacy of international law.[35] Unlike terrorism, the method is to break the laws like bread, sacramentally. It is to bring about changes in public attitudes that will influence governments and may even change them through the ballot box, but is not aimed to replace the electoral process.

Two Questions and Some Grounding

In what follows I want to focus on two questions. First, how was Hallam's, and therefore during its period of exponential growth, Extinction Rebellion's wider understanding of nonviolence understood? And second, how were Hallam and related movement influencers interpreting climate science to drive protest? Before I address them, a note on where I stand on a core concept. I have already made clear my personal positions – some might say biases – towards both consensus science and democracy that is constitutional, and, when on a large scale, primarily representative. Before proceeding further I must add a third, which is an explicitly spiritual understanding of the basis of nonviolence. By the 'spiritual' I mean partly the theologian Walter Wink's sense of the 'interiority' of outward things,[36] and also the profound inner interconnection of all things, the meanings of life as love made manifest. As the Vedic sages of Hinduism put it, our deepest nature is *sat, chit, ananda*. Being, consciousness and bliss. Self-realisation is the 'realisation' – becoming real in consciousness – of this great underlying truth, to which some may or may not choose to give such names as Goddess, Allah, Jah, Buddha Nature, Brahman, Christ, the Tao or God.

Can we square that with a love of and appreciation for science? I consider so, if we permit empirical scientific method to include the study of spiritual experience. The fact is that such perceived experiences do happen, either spontaneously or by practice within people's consciousness.[37] It has always been known in the East and other mystical traditions, and has been studied in the West at least since William James gave his seminal Gifford Lecture series on religious experience at Edinburgh University in 1901–2.[38] What are variously called peak, transcendent, cosmic and mystical experiences can arguably shed a fresh light on human nature and wider reality. Consciousness – that 'basic call' to awakening – cannot just be brushed aside. It is both experienced by us, and mediates every aspect of our known experience. To try and cut science off from this, as if it can float separately from experience, would be to try and separate the known from the knower's field of knowing. The quantum physicist Werner Heisenberg might have cast uncertainty on presuming any such a principle. These days, the field that studies such matters within the social sciences is known as transpersonal psychology.[39] If we are concerned about the deeper human implications of the climate change emergency for spiritual emergence in these times, it might reward at least a little study.[40]

My readers, however, might be relieved to know that detailed discussion of such scholarship is not one for this book. It is, however, important to point towards as a grounding for both the theory and practice of nonviolence that I share with many practitioners from differing traditions. They include Mohandas Gandhi, Peace Pilgrim, Daniel Berrigan, Bādshāh Khān, Thich Nhat Hanh and Leymah Gbowee, the 2011 Nobel Peace Prize joint-laureate of Liberia. In this matter, I respect but differ from writers like Gene Sharp, the influential author of *The Politics of Nonviolent Action*, who takes an explicitly instrumentalist (or utilitarian) view of nonviolence that can be dismissive of the spiritual dynamics;[41] or Chenoweth and Stephan's book that, for all its scholarly value, avoids discussion of the spiritual underpinning. What I have said here is my personal position and not one with which I expect others to accord. But it does make a fundamental difference in how nonviolence is understood to work and perhaps, in its long-term outcomes.

Satya or Asatya in Nonviolence?

The first of my two questions – how nonviolence was being understood – applies not just to Hallam and Extinction Rebellion, but also to wider activist and climate advocacy contexts. We saw how the theory around engaging 3.5 per cent of the population in rebellion to achieve revolution rests on a single point source: the book by Chenoweth and Stephan. However, as the investigative journalist Nafeez Ahmed points out, such research is based on the overthrow of *tyrannical* regimes. 'None of them involved successful nonviolent efforts to overthrow or change a Western liberal democracy.'[42]

One riposte is to say that Western liberal democracies are not true democracies. I partly agree with that given inadequate control over moneyed and media political influence. Studies suggest that in some countries, Scandinavian ones especially, most people are satisfied with their democracies, but in others and in a rising number, including the USA and UK, there is dissatisfaction. The UK has seen dissatisfaction rise from about 40 per cent in 2017 to about 60 per cent by 2020. What makes the difference between contentment and the malcontents seems to be the four 'Ps' – polarisation, paralysis, perfidy (or scandal) and powerlessness.[43] To me, this mandates reform and regulation rather than seeking to overthrow the system, unless there is a clear majority public buy-in to do so. As such, I consider that Hallam has made a 'category error' in his application of the 3.5 per cent doctrine. Most liberal democracies are not made up of populations that are critically poised, just waiting for the right trigger to overthrow their tyrant. Most people opt for political evolution, not rebellion and revolution.

But my problem runs much deeper than arguing about the numbers. In both *Common Sense for the 21st Century* and a key Extinction Rebellion teaching video, Hallam's approach to nonviolence is instrumentalist.[44] No doubt he has a deeper side to him, but what comes over is a means-to-an-end pragmatism. While he draws tactically on the lives of Gandhi and Martin Luther King Jr, he neglects their inner source of spiritual legitimacy, vision and resourcing from a place beyond ego. Gandhi was explicit on the matter. Instrumentalism was out. 'Non-violence is not a garment to be put on and off at will. Its seat is in the heart, and must be an inseparable part of our very being.'[45] Anything less, any reduction to the calculus of a numbers game, and we only reproduce the domination systems to which we thought we were opposed, and which substantially comprise the root drivers in human behaviour that result in climate change.

This distinction between instrumental and spiritual understandings of nonviolence is easily overlooked because, as Sharp, Chenoweth, Stephan and others have all shown, both approaches can 'work'. Does the distinction therefore matter? Is it not better to hide the spiritual card if we have one, and thereby muster all the more support onto the streets? I believe that if we are to escape the hubris of power, our relationship to truth must hinge on something close in ethos to what Gandhi understood. A certain spiritual teacher who he admired once said: 'Blessed are the meek, for they shall inherit the earth.' It sounds a bit naff, until you look at the original Greek from which 'meek' is translated. English has no single word equivalent for *praus*, which means the combination of tenderness and strength.[46] Only then do we see the power of such an ecological talisman: 'Blessed are the *gentle strong*, for they shall inherit the earth.'

As the spiritual figurehead of India's independence, Gandhi employed the Sanskrit word *ahimsa* for nonviolence. It means literally, 'not striking', and to the driving force behind this capacity he gave the name, *satyagraha*. This draws from two Sanskrit roots.

Satya, meaning truth, being, reality, essence, soul or God, for such terms flow into one another in the mystical mind of Vedic philosophy. And *graha*, meaning insistence, holding or force. *Satyagraha* is therefore variously translated as truth force, soul force or reality force. This has amazing implications for a principle like Extinction Rebellion's 'tell the truth'. The Mahatma, or 'great soul' as Gandhi was called, drew his inspiration both from the Vedic scriptures and from Jesus' insistence on the same core principles. Although a Hindu; indeed, probably because he was a fully engaged Hindu, he understood the unity and real presence of what Christ called, 'the way, the truth and the life'.[47] He wrote, and this passage is pivotal:

> The world rests upon the bedrock of *satya* or truth. *Asatya*, meaning untruth, also means nonexistence, and *satya* or truth also means that which is. If untruth does not so much as exist, its victory is out of the question. And truth being that which is, can never be destroyed. This is the doctrine of *satyagraha* in a nutshell.[48]

'Tell the truth' therefore cannot be for governments alone. It can never be a demand projected only onto others. The hand that points may point other fingers back towards the holder. To suggest that our predicament over climate change arises purely out of being the helpless victims of external oppressive forces, is to deny both our share in responsibility – however much constrained by options available to us and by misinformation – and our capacity for agency, the capacity to act. Gandhi said that if we start to understand ourselves as souls, we are never left bereft of agency. It is rooted in our dignity upheld by God. That is our sword and shield, the force of *graha* that emanates from *satya*.

Such spiritual reasoning, nay such experience, is unacceptable and invisible to the materialistic worldview. It is of a deeper interpenetrating order of being. It is why the Mahatma repeatedly insisted that 'The root of *satyagraha* is in prayer.'[49] Prayer is more than any form of words. Prayer is letting go into what Sanskrit calls the *atman*, the great Meridional Overturning Circulation of the soul. A space from where we not so much *demand* of others – for that would violate their dignity – as *beseech* of them. To *seech* or 'seek', as lovers seek each other out. Then with dignity preserved we have a coupling point from which to build conciliation. Then we start to glimpse what Quakers understand as 'that of God in all'.

Any lesser vision of humanity would becloud our very consciousness – *sat, chit, ananda* – being, consciousness, bliss. But, as the expenses-claim accountant would shake their head and say, 'no chitty, no kitty.' No supporting documentation, no reimbursement. No consciousness, no nothing else. It would draw us out of *satya* and into *asatya*; out of truth and into untruth; out of reality and into unreality; and out of nonviolence and into violence. That is why the question – How is nonviolence understood? – matters more than ever might first meet the eye.

Extinction and its Ripostes

Which brings us to my second question, around the interpretation of science used in driving activism. When Extinction Rebellion began, it conveyed a sense of being *witnesses* to the cascade of plant and animal extinctions that are escalating around the world as many habitats become less habitable. There is no scientific quibble with that. However, the narrative soon escalated to human death on a massive imminent scale and, as we saw Hallam putting it: 'How do we avoid extinction?'

His *Common Sense* manifesto bases his views on 'one recent scientific opinion', warning of 6 to 7 billion people dead as a result of climate change 'within the next generation or two'. The paper cited as his authority in the footnotes makes no such claim.[50] This is purely Hallam's extrapolation of a 5°C world given what *Common Sense* calls 'the central role of methane in a climate emergency . . . with the system spiralling out of our control and the likelihood of global collapse within a decade or two'. He reiterated this dieback claim in a BBC News interview feature, trenchantly insisting: 'I am talking about the slaughter, death and starvation of 6 billion people this century – that's what the science predicts.'[51]

Climate Feedback, a website more used to taking on deniers than alarmists, invited an expert panel to give their opinions on this prediction. The responses ranged from 'an illustration of a worst-case scenario' to 'wild speculation'. Ken Caldeira, senior scientist at the Carnegie Institution, put it bluntly: 'I know of no climate model simulation or analysis in the quality peer-reviewed literature that provides any indication' that there is a substantial probability, above zero, of 6 billion deaths this century.[52] However, Roger Hallam has not been alone in making such forecasts. James Lovelock, of Gaia-hypothesis fame, said in 2008 that he expected 80 per cent of the human population to be wiped out by 2100.[53] Four years later, he conceded that he had been 'alarmist' – that was his word for it – and had been 'extrapolating too far'.[54]

Those senior activist figures who have made related claims have run up against sharp pushback from the scientific community. Rupert Read is a spokesperson for Extinction Rebellion and well respected for his work on green philosophy. However, at the Schools Climate Conference in 2019 he climbed up on top of the rostrum desk for emphasis, and announced to a theatre of some 200 early-teenage children that it might no longer be *what* they are going to do when they grow up, but 'what are you going to do *if* you grow up'. The incident was reported in *New Scientist* as well as through less specialised news outlets because several climate scientists expressed alarm.[55] These included Dr Tamsin Edwards. As one clearly in sympathy with Extinction Rebellion as well as being an IPCC lead author who specialises in the effects of Greenland and Antarctic ice on sea levels, she tweeted out a video of Read's lecture, with the plea:

> Rupert, I am shocked at this talk. Please stop telling children they may not grow up due to climate change. It is WRONG and ... with these kind of statements you undo all the hard work of the scientists and pro-science XR people who I know are trying to keep to the (complex) evidence base, and be clear when they are citing outlier or extreme predictions.[56]

I asked Read why he takes such positions. He said that he doesn't think that climate scientists, each with their own specialisms, have sufficiently understood the 'fragility' of crop production when exposed to potential cascading consequences of climate change, and how this could quickly lead to social breakdown. However, SRCCL in particular makes a number of such references, including stating that more research is needed into 'the cascading impacts of land, climate and food security and ecosystem service interactions through different domains such as health, livelihoods, and infrastructure, especially in relation to non-linear and tipping-point changes in natural and human systems'. In a telephone conversation together, we agreed that while there are many parallel threads in our thought, there are some significant differences, and these we have aired with convivial disagreement in a subsequent Zoom symposium.[57]

One who would concur with Read on this point, and push it further, is Jem Bendell, a business school sustainability professor at the University of Cumbria in the north of England. An expert in digital currencies, his staff web page playfully describes how it earned him the moniker 'Professor Bitcoin'.[58] Bendell's contribution to Extinction Rebellion's *This Is Not a Drill* tells that he 'grieved how I may not grow old'.[59] The manifesto thesis for which he is now known, *Deep Adaptation*, anticipates 'inevitable near-term social collapse due to climate change' resulting in 'probable catastrophe and possible extinction'.[60] This, as he wrote on his blog, could be expected 'in many, perhaps most, countries of the world . . . within 10 years'.[61] He spelt out both the imminence and what it would look like in a roundup of where he considered the climate science stood as of 2018:

> But when I say starvation, destruction, migration, disease and war, I mean in your own life. With the power down, soon you wouldn't have water coming out of your tap. You will depend on your neighbours for food and some warmth. You will become malnourished. You won't know whether to stay or go. You will fear being violently killed before starving to death.[62]

Deep Adaptation was originally an academic paper that had failed peer review for lack of scholarly rigour. Bendell posted it to the web in 2018, achieving an astonishing half a million downloads within the first year. Part of his rationale leans on what he describes as 'data published by scientists from the Arctic News'. However, Arctic News is no scholarly tome. It is a blog site boasting over 7 million page hits that, amidst lurid illustrations, invokes the methane bomb and projects a possible global temperature rise of 10°C, by 2026, based on 'adjusted NASA data' heralding the 'mass extinction of man'.[63] Again, the pushback comes from within the scientific community itself. A journalist asked Gavin Schmidt, the director of NASA's Goddard Institute for Space Studies and one of the world's leading climate experts what he made of Bendell's paper. Schmidt said, and further pressed the point on his Twitter account, that it mixes 'both valid points and unjustified statements throughout', but is 'not based on anything real'.[64]

In a 2019 blog, Bendell responded to criticisms of his slant on the science. He describes his grief at having chosen not to have children, partly because they are 'the greatest contribution to carbon emissions that you could make' and partly out of 'the realization of the world they will have to live and die within'. He concludes that in future he will not be replying to, but rather, stepping away from, such controversies around his scientific claims to focus instead on building up the community around *Deep Adaptation*,[65] the activities of which include workshops, trainings, residencies in Bali, and an annual retreat at a yoga centre in Greece to 'support peaceful empowered surrender to our predicament, where action can arise from an engaged love of humanity and nature, rather than redundant stories of worth and purpose'.[66]

However, within a year of his withdrawal from scientific debate, he wrote a further blog having requested Schmidt to render his criticism specific. Schmidt obliged, providing a raft of reproofs including his assessment that *Deep Adaptation*'s take on Arctic methane was 'totally misleading', and that its pitch on runaway climate change was 'nonsense'. However, the professor, whose day job was to teach 'a sustainability-themed MBA programme', was unwilling to concede any significant ground to NASA's top climate scientist. Digging in his heels, the blog concluded: 'I have identi-fied two minor corrections and two clarifications I will make on the paper. However, none of those are material to the situation we are in and none of the main points are revoked.'[67]

Shortly afterwards, BBC News ran a feature that profiled Bendell and his most ardent 'followers' as 'climate doomers'. It quoted Myles Allen, professor of geosystem science at the University of Oxford, as saying that he considers *Deep Adaptation* to display 'the level of science of the anti-vax campaign'.[68] In counterpoint, it also cited Will Steffen, a retired scientist who had served on the Australian Climate Commission, suggesting that Bendell may be 'ahead of the game in warning us about what we might need to prepare for'. The pity of it all is that Bendell's core agenda – about the need for *resilience, relinquishment, restoration*, and recently he has added *reconciliation* – is both necessary and inspiring. That is why he has gathered such a following amongst people who are hungry for deeper meaning. We need people who, at their most effective, and if they discipline themselves to the settled science, can take an overview of things, drawing out what most matters, contextualising it and presenting it to the public in ways more digestible than the raw IPCC reports. There is for each one of us so much that is good and right to do anyway, without having to overreach our fields of expertise, conflate climate change with other causes and play fast and loose with signs seen in the sky.

Meanwhile, Arctic News' chosen doomsday date of 2026 doubles as the apocalyptic year of choice of Guy McPherson, a retired professor of evolutionary and resource ecology at the University of Arizona, and Bendell's referenced source in *Deep Adaptation* where discussing fears of an 'inevitable methane release ... leading to the extinction of the human race'.[69] McPherson, in turn and in a way that starts to feel rather circular, references his claims back to material from Arctic News, as well as to extrapolation from a range of scientific papers and other sources that, he says, 'even 10-year-olds understand ... and [that] Wikipedia accepts [as] the evidence for near-term human extinction'. The phrase used there, Near Term Human Extinction, has gathered a considerable ecopopulist cult following, complete with the social media hashtag #NTHE and online mental health support groups for the depressed and suicidal. The professor crisply reiterated and summed up his position in an interview given in 2018: 'Specifically, I predict that there will be no humans on Earth by 2026, based on projections of near-term planetary temperature rise and the demise of myriad species that support our own existence.'[70]

His website, Nature Bats Last, prominently offers suicide advice on its home page. While advising against such a move, he counsels that it can nevertheless 'be a thoughtful decision', and with this endorsement he bizarrely links to the post-mortem website of Martin Manley of Kansas, who intricately blogged the preparations for his own departure by gunshot in a parking lot.[71]

My first question was about Hallam's and related influencers' understanding of nonviolence. My second, about their interpreting climate science in ways that amplify people's fears beyond the expert consensus evidence base. For those who believe in the severity and particularly the imminence of their prognostications, such alarmism arguably crosses over into the realm of fantasy. If conflated with reality, this risks its own potentially tragic consequences.

Alarming versus Alarmist

There are other sides to the position that I have taken against alarmism. An activist friend put it to me that what Bendell's work does – and the same could be said of Read's – is that it pushes a point to make a point. It usefully brings people to the state of breakdown, from where they can break through into the new social norms that are demanded by deep adaptation. Read, in what he calls a 'friendly critique' of Bendell's work, speaks of the need to build a 'lifeboat civilisation'; a base from which a new beginning can be made.[72] His work emphasises the precautionary principle – the maxim that we shouldn't continue to do something harmful, like ratcheting up the carbon count, when we don't know or can't contain all the risks. This makes ecological sense, but risks must be responded to and not just reacted to, otherwise we'd never cross the road.

My view is that if a case can't be made without it being over-egged, either the case is not valid or those to whom it is being pitched are being spun. Exaggeration or invoking fear and panic only entrenches positions and sets up a backlash. For the purposes of motivating action, if anything is going to be enough, the unembellished science is quite bad enough to be good enough.

There is a longstanding debate in social psychology as to the sustained effectiveness of fear. In what is now a textbook standard experiment from 1953, psychologists Janis and Feshbach set out to test the effect of pictures that induced a weak, moderate or strong fear of tooth decay in a group of children. Questionnaires administered immediately afterwards showed the effect that many might expect. A week later, however, this reversed, and it was the least fearful but most informative imagery that left a more lasting impact on tooth-brushing behaviour. The psychologists concluded that the most shocking pictures of mouthfuls of rotting teeth led to a defensive resistance and avoidance response, such that: 'The over-all effectiveness of a persuasive communication will tend to be reduced by the use of a strong fear appeal, if it evokes a high degree of emotional tension without adequately satisfying the need for reassurance.'[73]

I get people coming up at my talks, or sending in an email, then being disappointed when I tell them that I only partly buy into the fears stimulated by prominent alarmists. Because I say I'm sticking to consensus science – even knowing that it can never be bang up to date and that its expression will be sure but probably cautious – I suspect they sometimes think that I'm the denier. A climate model researcher in Sweden dropped me a line, saying that he gets the same disappointed reactions, adding that 'some teenagers are distraught on this, so the alarmism of such actors is taking a heavy and unjustifiable psychological toll on others'.

Those who work with young people warn of the consequences of growing 'climate anxiety'.[74] Depending on its traumatic depth and the stimuli that sustain it, fear doesn't always have such a short half-life as Janis and Feshbach suggested. My friend Matt Carmichael, a teacher in Leeds who advises other educators on the implications of the climate crisis, said in an email that children's mental health is already fragile. Though himself a veteran of climate protests, he finds alarmism 'callous'. Those who pump it out in the absence of 'reassurance' such as the dental experiment spoken of – or, I would venture to extrapolate, in the deficit of avenues to express meaningful agency – don't see the consequences that teachers and parents later have to catch. Most children have not yet learned how narratives work and are constructed, and how to deconstruct and critique their premises, logic, conclusions, motivations and wider framings. Matt's message concluded: 'It's going to be absolutely essential for schools to learn to stick to the straight and narrow of mainstream science and shun anyone who thinks they know better.'

To me, it feels as if a kind of projective anxiety is often being played out where climate change becomes a lightning rod both for the anxieties that it generates, and for other anxieties that people carry in these dislocated times. It might be said that even to suggest such a thing is dangerous. The deniers will quote it back, and then what? But I'm more interested in exploring truths. How else will we earn the respect of those with whom we disagree but need to reach? When I listen to the way that frightened people talk about global warming it can seem as if the climate can become what psychotherapists call a 'chosen trauma' – a focus of meaning that objectifies and seems to make sense of wider constellations of anxiety in their lives.[75] Such misattribution (if that it is) overloads climate change. By turning it into a fetish in the anthropological sense – something that becomes emotionally magnetised beyond itself – it can grow into a scapegoat onto which a range of compound personal and social problems are discharged. If this view has any validity, then merely cutting CO_2 emissions, while vital in its own right, is not going to take away the underlying causes of malaise, including those that drive consumerism at the cutting edge of climate change. Also, as climate scientists in the public domain are finding out, they're going to need the skills that draw not just from physics, but from realms more metaphysical.

None of this is to suggest that what is happening to the planet ought not provoke anxiety. On the contrary, in the early stages of writing this book I had several disturbing dreams. One was of a well-known climate activist whose assassination on his bicycle I watched played out in slow motion. There was absolutely nothing I could do to stop him cycling at a manic speed, head down, round a corner, up a steep hill and blindly into a volley of automatic gunfire from a special forces hit squad, two men dressed all in black, who were waiting in ambush on a powerful motorcycle. In another, I was swimming through the filthy flood waters of the River Clyde that were rushing through the streets of a medieval part of Glasgow. This time I had agency. I was equipped with a mask, snorkel and flippers, and was able both to dive and to show others how to swim across to safety, making use of recirculating eddy currents such as one learns to spot and ride when out canoeing. Taking both dreams together, as digested later from my notebooks, it's as if they said we don't have to pedal grimly onwards into death. If we look out for the eddies, there are ways and means that we can learn to navigate. Often at such times while writing, I held on, as if they were a talisman, to T. S. Eliot's words: 'Not fare well, / But fare forward, voyagers.'[76] Life is not a journey from which we bid 'fare-well' or can expect to fare comfortably. This is one on which we fare as life will have it unfold, voyagers.

I said to the atmospheric astrophysicist Katharine Hayhoe, that I often find myself racked between the deniers and the alarmists, trying to hold on to the humanity of both, recognising their fears or differing priorities, and yet insisting on consensus science. She answered, 'It is a narrow and lonely place so it's great to have company!'[77] The climate physicist Michael Mann concurs. He sees 'doomism and despair' that exceeds the science as being 'extremely destructive and extremely influential'. It has built up 'a huge number of followers and it has been exploited and co-opted by the forces of denial and delay'.[78] 'Good scientists aren't alarmists,' he insists. 'Our message may be – and in fact is – alarming . . . The distinction is so very, very critical and cannot be brushed under the rug.'[79]

Neither Hayhoe nor Mann are the kind of scientists who take distance from campaigning as 'climate advocates', as the former puts it. Both openly support and encourage protest that rests on a firm evidence base. In April 2019, they were amongst the twenty-two lead authors of a letter to the journal *Science*, headed 'Concerns of young protesters are justified.' The prominent English climatologist, Kevin Anderson, who is an advisor to Greta Thunberg, was also one of the lead authors. Along with more than 3,000 other experts who added their names as co-signatories, it stated: 'We call for our colleagues across all disciplines and from the entire world to support these young climate protesters. We declare: Their concerns are justified and supported by the best available science.'

A further point was put to me by John Ashton, now an independent advocate but formerly the UK climate change envoy and a seasoned diplomat. Extinction Rebellion and (especially) Greta Thunberg had, he said, changed the game, forcing up the urgency and ambition of the political response. He made the observation that time in politics is binary. The only categories are today and tomorrow. And the only problems that get serious attention, as opposed to lip service, are those labelled 'today'. The accomplishment of the new climate activism has (for now, and only in some societies) been to make the climate a 'today' problem for the first time. The voice of science, in the IPCC and elsewhere, has been essential in putting the issue onto the agenda. But as a political signal it has been too weak to drive the transformation that the findings of the science demand.

The tension, then, is not between science and protest. The tension is between science and multiplying up its extreme ends of likelihood in ways that are tantamount to pseudoscience: 'If the worst imaginable happens it is this. And if the worst of that happens, it is this.' The ancient Celts may have been justified in their greatest fear that the sky would fall in. The asteroid may be on its way right now. But real science balances up the probabilities.

Nemesis, *Heiros* and Leadership

On 20 November 2019, while promoting the newly published book edition of *Common Sense for the 21st Century*, Roger Hallam gave an interview to the German weekly newspaper *Die Zeit*. In comparing climate change with historic genocides, he described the Jewish Holocaust as 'just another fuckery in human history'.[80] In another interview for *Der Spiegel*, he defended his billions-dead claim on the basis of 'predictions by the IPCC', telling the journalist 'go and do the math' and that 'climate change is just the tubes that the gas comes down in the gas chamber'.[81] A third interview emerged in

Frankfurter Allgemeine where, in referencing the number of Jews killed, he was reported as saying, 'Who gives a fuck about numbers, right?'[82]

German Extinction Rebellion groups expressed their utter devastation at how such a devil-may-care relegation of the Holocaust would pander to the far right in their country's highly polarised politics.[83] Hallam apologised 'for the words I used'.[84] But any hope of the storm blowing over disappeared when it emerged that he had premeditated the incident, instrumentally, as a provocation to generate publicity, and had written a memo to colleagues in the Media & Messaging team of Extinction Rebellion UK that laid out a strategy of 'designing this sort of "trap" for the media'.[85]

Extinction Rebellion Scotland formally dissociated from him for 'language that dog whistles far right wing extremists'.[86] George Monbiot tweeted 'with great sorrow' that it was essential for the movement to exclude him.[87] XR Jews called for anyone in future leadership positions to be given training in 'anti-oppressive practices and decolonisation, including training on modern anti-Semitism'.[88] An open letter from some of its Regenerative Cultures Circle Coordinators, including co-founder Simon Bramwell, acknowledged Hallam's passion and dedication in everything he does, but said that 'founder's syndrome' had given rise to 'an end-justifies-the-means doctrine' that led to inflexible fundamentalism.[89] Much to its credit, Extinction Rebellion UK set in place a restorative process looking at questions of responsibility, autonomy, decentralisation 'and how power is used'.[90]

Hallam agreed to step back out of the limelight while Extinction Rebellion worked on its process, but this agreement didn't hold for long. By March 2020, hopes of a satisfactory outcome had stalled. Hallam took off in his own direction, announcing a book promotion tour in America which left observers pondering how such travel squared with his drone protest at Heathrow. An official update on the restorative process stated: 'This trip has not been funded by or agreed with Extinction Rebellion UK. Roger is not a spokesperson (in person or in the media) for Extinction Rebellion.'[91] As it happened, and unsurprisingly given his prison sentence over a matter of airport security, the trip was thwarted by

US Homeland Security on visa grounds at the point when he was due to fly out. Accordingly, the tour commenced in virtual reality – by video link.[92]

Meanwhile, Rupert Read, writing with two colleagues in a personal capacity, took a leading role in Extinction Rebellion's forward strategy. Early 2020 saw the launch of a frank, heart-searching policy pamphlet addressed to 'all rebels'. *Rushing the Emergency, Rushing the Rebellion?* distanced itself from Hallam's 'overly polarising tactics' and questioned the 3.5 per cent theory of change.[93] It stated that the organisation's own YouGov polling showed that public support had fallen to just 5 per cent, down about thirty points from its peak traction in October 2019.[94] Donations were down, so were social media interactions, and it conceded that this, 'leads us to conclude that XR is already approaching terminal fatigue with its existing methods'.

Read's proposed alternative way forward was to highlight the public's vulnerability to systemic collapse. Both in the pamphlet's text, and more pointedly so, in recorded talks given around the same time, he proposed disrupting the just-in-time delivery system at supermarket distribution centres. With a thousand or so activists on the case, such blockades would bring about not malnutrition or starvation, but a temporary shortage of such foods, he suggested, as rice or pizza. Water, drought or flooding could also become a focus for 'rebels', with the effect that '. . . you could imagine a kind of action that would mimic those sort of things . . . to create a sort of miniature but real sense of local and international crisis.'[95]

Such actions, said the pamphlet, would be 'a smart wake-up call'. They would be '. . . something beautiful, powerful, intelligent, meaningful, and resonant, that challenges the authorities to either arrest you *en masse* – risking great public sympathy – or lets you get away with it'. And they would be 'a route by which we can arrive at numbers so huge, and popular sentiment sufficiently supportive, that mass mobilisation in the capital will be overwhelming.'

Even as I quote those lines, I can feel the discomfort of movement activists I know who have sacrificed much to make their point for the common good. Quite apart from its sense of public messaging, it leaves unanswered an obvious question. If exposing vulnerability would have such a profound effect on public consciousness, why has this not already happened? Say, from the 1966 seamen's strike state of emergency,[96] the supermarket shortages of the 2000 fuel tax protests and panic buying during the 2008 banking crisis? I know Rupert from other contexts and spoke with him, just as COVID-19 was ramping up and America had blocked flights to Europe. I questioned his presumption of public sympathy. He said, 'depends how it's done', noting that the upheaval caused by the virus had rendered the question irrelevant 'for the moment'. He confirmed this shortly afterwards, saying that it was 'not necessary anymore' because the pandemic had achieved the same awareness raising.[97] Time would tell whether that would translate over to climate change.

In such 'end justifies the means' tactics, as deployed by Roger Hallam and skirted close to here by Rupert Read, we might see the nemesis of instrumental uses of nonviolence. If we might take the supermarket protest as originally proposed and make a contrast: when Gandhi marched to make salt at the seashore in defiance of the British Raj, he never sought to deprive others of their salt to show them what it felt like. His engagement of the powers that be was at a more subtle level, one that exposed their bankruptcy of moral authority. Just where such autonomy of individual leadership in debatable directions leaves questions of governance and reputation in Extinction Rebellion as an organisation managed by 'holacracy' – an approach that allows for a high degree of creative freedom in self-organising work or core groups – is a question that many other radical movements might also ask.[98] Suffice to say that all of us are works in progress. Where Hallam fell, and where Read in my view risked opprobrium, we too, in other contexts, might have stumbled had we not first heard the warning cries ahead.

There is a long-running tension here of leadership in groups that seek to be leaderless. Many good people in many good organisations and movements are wrestling with such issues of power and accountability. Whilst I may be critical of some key aspects of their work, I would trust these to include, at least in their best intent, those such as Hallam, Bendell and Read. The debate goes back at least fifty years. In 1970, the feminist advocate Jo Freeman wrote a now-classic essay called 'The Tyranny of Structurelessness'. She argued that there is no such thing as a leaderless group.[99] If power is not acknowledged and organised through what she called 'democratic structuring', governance is likely to fall under the control of founder figures or small circles of friends who comprise 'informal, unplanned, unselected, unresponsible elites – whether they intend to be so or not'. Then, as the adage has it: 'Power denied is power abused.' As such, whatever be our cause we end as 'rebels without a clue'. The counterculture's rejection of authoritarianism after the Second World War was desperately necessary, but it remains that daunting work in progress.

I am struck that the participative democracy of New England town meetings had its roots in religious traditions. Originally, church membership was usually required to be a participant.[100] Presbyterianism as a prominent influence elects 'presbyters' or elders from the community for leadership, and normally a congregation 'calls' its minister, which, in ways that are now antiquated, has nevertheless led it to being described as 'the seedbed of democracy'. Quakers use a form of decision-making and leadership based on spiritual discernment. When it works well, this aims to set ego concerns aside and collectively seek a deeper 'leading', 'prompting' or 'movement' of the Spirit.[101] Many alternative leadership forms and processes are rooted in Quaker practice including 'clearness meetings', 'sociocracy' and support groups – the latter often having origins that trace back to the deliciously quaint-sounding Quaker principle of 'meetings for sufferings' to uphold people engaged in exceptionally exercising work.[102] However, the question always hovers in my mind as to how well such approaches, developed within explicit spiritual frameworks and holding, can translate to secular contexts. Spiritual leadership is based on the principle of 'servant leadership', by which 'the first shall be last and the servant of all'[103] – but such humility can sit uneasily in the modern world.

I would press the question further by observing that the term 'hierarchy' comes from the Greek, *hieros* ('sacred') and *arkhein* ('order'). As such, the notion of it meaning a top down authoritarian structure of command and control applies only if one's sense of the 'holy' is of domination from above. But if one's sense of the holy – of ultimate concern – emerges from within, then a very different insight opens. Then hierarchy can equate with 'development' in the endogenous sense that we discussed earlier. Then power becomes not power over but empowerment from within, through the shared discernment of right ordering. Then, as we will later see, it all equates with ancient concepts of a set-in-place and sustained reality that find expression in such Eastern-rooted terms as *satya*, *dharma* and our curious English word, 'doom'.

It follows, also, that the *authoritarian* as perverted hierarchy must not be confused with the *authoritative*, which should be *authored* out of *authenticity*. Authentic to what? Authentic to the Spirit that gives life. And if there's no such spirit? Then we're stuffed anyway. And if there is? Then, as we saw at the end of Chapter 3 with the ecology of Aldo Leopold, and as a statement that should not be applied rigidly: 'A thing is right when it tends to preserve the integrity, stability, and beauty of the biotic community. It is wrong when it tends otherwise.'

Climate Millenarianism

My appeal, therefore, and my reason for bringing my reader to this point, is to consider not taking the *heiros* out of the *arkhein*. I do believe that to not allow an opening to the possibility of spiritual reality can leave people very vulnerable. The momentous times in which we find ourselves today are ripe for what social psychologists and anthropologists describe as 'millenarian' movements to emerge: 'Become one of us and let go your angst, do this and that, make sacrifices and follow the leader – and a new millennium will be born, where the wicked bygone order will collapse, and an elect world order of the redeemed will be ushered in.' I do not think that Extinction Rebellion is a 'cult' as some critics have suggested, but, to my ear, Hallam's *Common Sense* and other apocalyptic writings on climate change that obsessively exceed the science do set a slight alarm ringing in the background.

In his classic 1957 study – again, from the pioneering heyday of social psychology – Norman Cohn wrote that cults and millenarian movements picture salvation as:[104]

Collective – to be enjoyed by the faithful as a collectivity;
Terrestrial – to be realised on this earth;
Imminent – to come both soon and suddenly;
Total – to utterly transform life on earth [as] no mere
     improvement on the present but perfection itself;
Miraculous – to be accomplished by, or with the help of,
     supernatural agencies.

The first two of these are necessarily germane to the nature of climate mitigation and adaptation. The third is partly germane, but exceeds its apocalyptic remit if the degree of imminence is exaggerated. The fourth is utopian, edging towards totalitarian. The fifth, in my opinion, if uprooted and secularised from a deeper theological context, is pure magical thinking.

Surely the fifth is not in anybody's ballpark? Well, there's my problem with a dogged insistence on net zero 2025. If it was only there as a political stalking horse, a biting spur to action, then fair enough; but tell the truth. If, however, 2025 was put there in 2018 because people really believed that it was both technically and politically achievable, a 'demand' that could reasonably by mass protest be thrust upon government, well, I may have very blinkered vision, but that sounds magical to me.

The Possibilities of the Future

A very deep wisdom – partly ancient and partly emergent – is required to see us through the explosive effects of something as huge and as global as the climate crisis. Irrespective of what is or isn't happening in the Arctic, a methane bomb goes off inside the mind. Fear – whether justified, manufactured or indeterminate – activates archetypal processes in the psyche at both individual and collective levels of being. Liberation theology speaks of the 'irruption' of both the poor and the spirit – a bursting in of the holy, as distinct from an 'eruption' that bursts out.[105] Many in Extinction Rebellion, and increasingly in social and environmental movements more widely, have understood or are coming to understand this irruptive potential. However, such understanding, such a bursting in (paradoxically, from within) of new potential and ways of being, needs to extend further into climate activism and more formal advocacy so as to deepen its grounding. Our work needs to come not just from the materialistic and rational perspectives of good science, but to be a spiritual activism. In so saying, it is crucial to study and discern what distinguishes the authentic from the inauthentic, the true liminal from the false liminal, the holy from the phony holy. This is what a spiritual tradition at its best, which is something that must be sought for and discerned, can help people to do.

Properly understood from the Greek once again, a *charism* is a spiritual gift. Faith, hope and love (or 'charity') are all charisms. If residing in the truth of *satyagraha* – the *satya* as being, as God, as reality itself – then the prophetic or, more widely, the shamanic function will work in ways that open avenues of blessing.[106] Life can then flow back into a deadened community, including communities writ large as nations. But if awareness, surrender to the greater whole and perhaps the guiding support of elders in a spiritual tradition are lacking, charisma can instead inflate the ego.

Much of advanced spiritual teaching is concerned with recognising and overcoming this problem. Hinging on it is the distinction between the authentic, as that which remains rooted in the divine, and the cultic, as that which has detached from it. If spiritual gifts are not returned as service to the higher power or *graha*, then wittingly or unwittingly, and whether at individual or collective levels of agency, the gift that has been misappropriated will transmogrify. Recklessness will lead to wrecking.[107] As a *Hadith* or oral tradition of Islam has it, 'The most excellent *Jihad* (or spiritual struggle) is that for the conquest of self', in the sense of the 'small self' or ego.[108] When the ego is not grounded in the 'great self' of divine interconnection; when its insistence on the sovereignty of its own presumption of rationality obscures its sensitivity so that it will not even entertain intimations of the greater mystery, then power denied can become power abused in a very big way.

So it was that shortly before his death in 1961, Carl Jung wrote that modern humankind does not understand how much its rationalism, which has dangerously destroyed its mythic and spiritual values, has put it at the mercy of a psychological underworld. We are now, he said, 'paying the price for this break-up in world-wide disorientation and dissociation'.[109] Our problem today is not too much spirituality, but a dearth across society of what the humanistic psychologist Abraham Maslow called 'the farther reaches of human nature'.[110] We must listen for the music of a higher calling that can rise to our times, and consider whether to tone down the hammers of hubris that have deadened our inner ears.

Alarmism distorts our temporal horizons of what is possible. As the veteran Greenpeace campaigner Chris Rose suggests, such 'gloom picking' leads to 'solutions denial' that ramps up 'climate grief' that exploits the poorly informed.[111] In their panic, many of its key proponents advocate potentially disastrous fixes, the magic bullet of geoengineering especially. I agree with those who say: 'There isn't enough time.' And yet, the opposite of one great truth is very often another great truth. As an Arabic proverb puts it: 'Haste is the key to sorrow.' If our politics are deep green, we must pay attention to the fact that, already, nativist forms of ecofascism have drawn blood on the growing alt-right fringes of drawbridge environmentalism. The Oklahoma 'Unabomber' and the Christchurch mosque gunman both appealed to certain types of 'green' narrative in their manifestos.[112]

All this is why I walk along the ridge of Katharine Hayhoe's 'narrow and lonely place'. To over-egg the cake is like those terrorist alerts that remain forever high. Alarmists who extrapolate beyond sound evidence may be right, but if so, by the wrong process. The upside, is that they may perversely hit it lucky and warn of something of which others had been too cautious. The downside is that in the long run they undermine the very principles of truth that they purport to speak.

Mike Hulme, a professor of geography at Cambridge University, was the founding director at the Tyndall Centre for Climate Change Research. He says, 'I am a denier. A human extinction denier,' and specifies: 'I will deny that there is warrant to collapse the possibilities of the future to human extinction.'[113] That phrase, *to collapse the possibilities of the future*, is pivotal. How so?

Because to do so feeds upon the natural fears and decent trust of the understandably uninformed. It then allows the enemies of climate action to paint climate science as the domain of wacky prophets and their followers, who have to keep on revising upwards their forecast date of doomsday. It draws those who have been caught up in such thinking into the cognitive dissonance reduction of looking for, and in a strange way maybe even hoping, that the signs on which they have staked so much are being fulfilled. The only remedy for such situations is that in our understandable despair and burning yearning for action, we must keep the head engaged, as well as the heart and hand.

Climate change denial is a waste of time. But climate change alarmism is a theft of time. We have no mandate to collapse the possibilities of the future, to contract and restrict our latitude for agency and action.

TO REGENERATE THE EARTH

We have now looked at where the science stands. We have looked at both ends of the spectrum, from denialism to alarmism, over which this plays out in public discourse. But what about the vast mid-range of that spectrum? What is happening in the mainstream arena around climate policy and action?

In this chapter, I will attempt to sketch some such ground. Do skim over it if matters like green new deals or the machinations of corporate social responsibility versus wider capitalism are not for you. If it is for you, I will touch on the promises and limitations of technology, followed by the responses of key segments of society: the *public and voluntary* mainstream, the *private sector* and the *vernacular economy* of grassroots initiatives. In raising the latter, I will merely lay a little philosophical ground to build upon in the next chapter's case study of community-empowered land trusts.

However, let me warn that I shall hit a dismal note in questioning whether the picture really adds up in the way that our world is currently configured. I shall therefore have to ask whether the gap between economic growth and ever-accumulating CO_2 emissions is just too great to bridge unless we also bite the unpalatable bullets of population and consumption. And with them, the values needed to approach them humanely, in ways that build on human rights, gender equality and justice for the marginalised.

Ecomodernism and the Rhythms of Life

In 1979 the counterculture magazine, *CoEvolution Quarterly*, published 'A Short History of America' by the cartoonist R. Crumb. His series of illustrations started with America as a wilderness, then the coming of the railroad, then the village with the telegraph, then the rustic town with electricity poles, and finally, in a relentlessly unfolding sequence, the gas-guzzling concreted-over city. A follow-up poster became a best-seller, with its three additional concluding scenarios. The 'worst case' shows the city ruined and abandoned in apocalypse. The 'fun future' depicts 'techno-fix on the march', with sci-fi buildings and those magnificent men in their gonzo machines. And the 'ecotopian solution', where convivial homesteads nestle into leafy neighbourhoods, and the only remnant whiff of reefer madness is of gentle hippie folks transporting kids on trailers drawn by bicycle.[1]

It surprised many when Stewart Brand, the magazine's editor, became in 2015 a signatory of *An Ecomodernist Manifesto*. Put forward by a group of leading futurists, this rejects the view 'that human societies must harmonize with nature to avoid economic and ecological collapse'.[2] Instead, as Brand also put it in his controversial book *Whole Earth Discipline*, 'We are as gods and HAVE to get good at it', a phrase that, interestingly, he adapts from the Psalms.[3] This means assuming an overarching technocratic control of nature by embracing advanced nuclear energy to satisfy growing world demand, GM crops that can withstand climatic extremes, a wholesale shift of population from the rural areas to efficiently run mega cities and the geoengineering of the planet's climate. In other words, fasten seat belts in your gonzo machine: technofix has lift-off, whether we've chosen such a trajectory or not.

Brand seems to take it for granted that the rural poor will fit their lives around the rich. 'The move to town is a liberation,' he writes, quoting with approval a modernising Indian politician who damned the village as a 'den of ignorance, narrow-mindedness and communalism'. The city, not the hamlet, is the seat of civilisation. For sure, slums are found pressed up against the affluent neighbourhoods, but that is 'mainly an efficient economic event typical of city density – service supply and service demand . . . the maids, nannies, gardeners, and security guards walk to work.'⁴ Much though I have been nourished by Brand's Jungian trickster role down the years – one that can trip us into wider realisations – there feels to this a little of the top-down view of Californian Man. As a young activist, Christina Waggaman, summed up in a parallel context: such men can come across as if they 'think they uniquely possess the knowledge or means to save the world'.⁵ It is certainly the case that rural life can breed small-mindedness, but urban life can breed expansive arrogance.

And yet, it has to be conceded that some aspects of the eco-modernist vision are already with us, and not without arguable benefits. For instance, in food production the massive productivity of high-tech agriculture shows what can be done in modified environments beneath plastic and glass. The Netherlands, although densely populated and with large areas of its territory below sea level, nevertheless exports huge quantities of food to neighbouring countries. Soil degraded to sand? No problem. The only productive function that it needs to serve is as a supporting medium for hydroponics. Nutrients can be drip-fed in liquid solution and, under such highly controlled environments, there's less need for such biocides as weed killers and insecticides. It will be true that there will be less space for the birds and the bees, but, goes the argument, that's compensated by freeing up inefficiently used land from conventional agriculture elsewhere.

It's not hard to step from this to imagining a world where, instead of heating greenhouses to grow food, rich countries that can afford the technology systematically cool them. And why not just go the whole hog, abandon earth and live in space colonies? What troubles me, is the disconnect encoded in such thinking between what can be a hard-driven urban pace and the rhythms of nature. I wonder what it does to our human nature? When a group of us invited the great Indian-Spanish Hindu-Christian philosopher Raimon Panikkar to speak in Govan in 1990, his theme was 'the rhythm of being' and from there to the qualities required for 'the survival of being'. He called his lecture, 'Agriculture, Technoculture or Human Culture?' And to illustrate how the human can get locked in to technological rhythms, he told a joke about the first ever Indian commercial airline pilot. After graduating from aviation school, he went back to the village to ask for his old mother's blessing. 'Now, my son,' she said as she clasped his hands. 'Flying is very dangerous. So promise me just one thing. Always fly very low and very slow.'

Ultimately, the city depends on ecological rhythms from the rural hinterland in a way that the rural hinterland does not symmetrically depend upon the city. The rural hosts the urban. But it's not an either/or. We need to hold the two in cross-fertilisation. The urban needs to understand what the rural contributes, and in human affairs, on a planet of 8 billion, rural dwellers need to be in tune with urban needs. More than that, as Panikkarji reminded in his Gifford Lectures from Edinburgh University, published just before his passing, there is the question of the greater framing. The *survival of being* as the foundation of all things. And yet, as he humbly concluded: 'How can human thinking grasp the destiny of life itself, when we are not its owners?'[6]

Green Technology: Hopes and Limits

But for now, let's get back into the realm of Logos rather than Mythos, to remain in the story rather than in its framing: another example of what is happening with advanced technology – of what gives grounds for optimism but also opens out new challenges – lies in developments in electricity production. Already the cost per unit of renewables-generated electricity has fallen to levels that were once thought unlikely before 2050. Where land or sea space is cheaply available, wind and solar farms are out-competing fossil fuels, and that's without an ongoing need for start-up subsidies. Short-haul electric or hydrogen ships are already being trialled. With advances in battery storage, Loganair in Scotland hopes to become the world's first airline to introduce an electric plane, testing it on short-hop flights between the Orkney islands as early as 2021.[7]

In many ways, the advances provided by science and technology are thrilling, yet silver bullets carry unintended impacts and

sometimes hidden costs. One example is that part of the justification for Britain's HS2 high-speed railway line was that it would cut
carbon emissions by reducing the need for so many flights from
regional airports to London. However, *Railway Technology* magazine, which is hardly a PR mouthpiece for the aviation industry,
reported that owing to the carbon footprint of the infrastructure
required, 'the new £60bn 185 mph London–Manchester rail line
could be less eco-friendly than the same air route over 60 years.'[8]

A similar concern hovers around electric vehicles because
green technology doesn't mean a zero carbon footprint unless all
the inputs are green. One study concluded that electric vehicles
will barely help to cut CO_2 emissions in Germany. A conflicting
German study however found that electric vehicles have lifespan
emissions 43 per cent lower than diesel equivalents.[9] Why such
divergences? Because so much depends on the assumptions made
about the carbon intensity of production (the batteries especially),
the source of the electricity used during the life cycle, and the
not-inconsiderable energy demands of end-of-life recycling. At
present, an electric vehicle in Germany currently runs some 40 per
cent on energy produced by coal and gas. A French one runs over
70 per cent on nuclear. An electric vehicle on the Isle of Lewis
would be blown away by the wind – were it not for the battery
providing ballast. In ecology, the ecosystem context matters.

Greening, like ecosystems themselves, is a systemic process.
You've got to work on all parts simultaneously, and upgrade the
system iteratively. After all, when Nikolaus Otto built the first
practical petrol engine in 1876, he'd never have imagined how large
a proportion of the world's economy would, within just a few decades, revolve around oil extraction, refining and delivery to retail
pumps. Today we've got to think the same about green electricity,
but not hide from the problems. A major issue is sourcing the
resources. The Natural History Museum points out that for the
UK alone to replace its 31.5 million cars on the road with battery
power, just under twice the world's present annual production of
cobalt would be needed.[10] Add in the need for other scarce min-
erals such as those from which 'rare earth' elements are extracted,
and troubling questions arise as to the impacts on front-line

communities perhaps displaced, polluted and violently threatened by the mining industry. For example, the shepherds of the Atacama Desert in Chile fear for their future as the battery boom opens up huge demand for the lithium mined there. What has grown into a billion dollar industry demands huge amounts of water for the extractive process. Its effect on water tables turns green pastures into desert.[11] Meanwhile, out of sight and out of mind, deep in the silent abyss, underwater sanctuaries will be wrecked as seabed mining starts to become a thing.

I feel a tightening in myself as I write that. On the one hand, is concern for nature free and wild. But the scale of what it can imply can provoke a backlash, as is evident in *Planet of the Humans*, the controversial documentary movie by Jeff Gibbs and Michael Moore. On the other hand is the background ticking of the carbon clock. To remind ourselves, by the early 1970s, the annual rate of global CO_2 increase was 1 ppm. By the turn of the millennium it was 2 ppm, and by 2016 it had topped 3 ppm and escalating. Metaphorically speaking, we've already eaten of the Tree of Knowledge. Where might lie a wise pathway between full-on eco-modernism and ecotopia? Can we learn to harmonise with nature and yet sustain high levels of population and consumption? Or do we need to think in other ways? I will round on this question in my last two chapters. They will be the strangest chapters in this book, but the main reason why I accepted the publisher's request to write it. For now, let us stay within the mainstream discourse.

Public and Voluntary Sectors: The Green New Deal

We don't have to replace 31.5 million cars on British roads with electric vehicles. That may be yesterday's solution to tomorrow's world. A rich summary of the general drift of the required new directions can be found in the *Science* letter mentioned earlier, the one that was written by scientists in support of youth climate protests:

> Policies are needed to ... make climate-friendly and sustainable action simple and cost-effective and make

climate-damaging action unattractive and expensive. Examples include effective CO_2 prices and regulations; cessation of subsidies for climate-damaging actions and products; efficiency standards; social innovations; and massive, directed investment in solutions such as renewable energy, cross-sector electrification, public transport infrastructure, and demand reduction [all with] a socially fair distribution of the costs and benefits . . .

A shopping list like this adds up to what is called the Green New Deal, a term borrowed from Franklin Roosevelt's 'New Deal' economic programme of social reforms and public works that lifted the USA out of the Great Depression of the 1930s. Environmental advocacy is not just about the 'push' of an 'obstructive programme' of protest. It is also, but usually carried out in back rooms far away from media cameras, about the 'pull' of Gandhi's 'constructive programme' for setting change in place. In this respect, Britain has a strong track record of the voluntary sector of civil society working closely with the public sector of government to come up with policies that have a hope of getting off the ground. One example is with the London-based Climate Coalition, which operates in alliance with Stop Climate Chaos Cymru in Wales and Stop Climate Chaos Scotland. Between them, these represent 140 organisations with a shared public membership of 22 million people.[12] Participants range from Greenpeace, WWF, Friends of the Earth and the National Trust, to the Energy Agency, Eco-Congregation churches, the Humanist Society and a scattering of trades unions.

A well-thought-through example of a new deal proposal from the voluntary sector is *The Common Home Plan* of the Scottish think tank, Common Weal.[13] It tackles requirements across sectors, including energy, transport, buildings, heating, food and land use. The report acknowledges that one size can't fit all, because every country's ecology varies. I like the honesty by which it is upfront about the costing difficulties, because '. . . the scale of the sums of money needed in some places is greatly out of line with the available knowledge'. It also concedes that while some sectors can be brought to zero carbon with relative ease – such as electricity generation and land-based transport – others, such as heating old buildings have very much more limited options. The price tag for its plan is £170 billion. For Scotland's 5.4 million population, that pans out at £31,500 per capita, or the cost of a daily pint in the pub for thirty years, or ten years if you happen to hang out in central London. Indeed, the authors claim that the costs are small compared with post-Second World War economic reconstruction. In addition, the massive boom in jobs would bolster tax revenues to help repay the debt.

A revealing feature of the *Common Home Plan* is to take it to the people, to carry it out and lay it down for debate and ownership in real-life communities, because 'we can only do this with the help of a lot of people.' Such public engagement gives expression to a key principle of Scottish thought called the democratic intellectual tradition.[14] This might merit some airing. The principle of the democratic intellect accepts that knowledge requires specialisation, and will therefore create elites. But that raises two problems. First, left to their own means, elites will tend to move in bubbles and to develop blind spots. Second, they can succumb to self-interest because knowledge can be power. The redress is also twofold. First, knowledge should be tested for its blind spots in the wider democratic body of the community – what is known colloquially in Scotland as 'the body of the kirk'. Second, knowledge should be of service to that community. Implicit to this principle is an old monastic notion that the benefits of education ought not to be appropriated primarily for private gain. Rather, they should be woven back into the fabric of the society that made them possible. It is on the commons of the community's social and environmental surfaces that all activity – including the freedoms granted to profit-making businesses – enjoys the privilege of taking place. Lest we forget.

The parallels will be evident here with approaches that owe much to a humanising trickle-up consciousness from countries of the south to those of the north. They cluster under such names as Farmer First, participatory appraisal, action research, appreciative enquiry and planning for real. They include Paulo Freire's principles of grassroots education from Brazil, as expressed in his book *Pedagogy of the Oppressed*. They connect also with Gandhi's sense that *satya* implies *swaraj*, or self-rule. This lead to his absolute emphasis on the soul-based dignity of every person, right up to our collective expressions as nations. As he put it:

> I will give you a talisman. Whenever you are in doubt . . . apply the following test. Recall the face of the poorest and the weakest person whom you may have seen. Ask yourself if the step you contemplate is going to be of any use to them. Will they gain anything by it? Will it restore them to control over their own life and destiny? In other words, will it lead to *swaraj* – freedom – for the hungry and spiritually starving millions? Then you will find your doubts and your self melt away.[15]

That's the principle of democratic intellectualism (and action) at its fullest. Looking at the lengths to which the IPCC these days goes in presenting several layers of depth to their major reports, we can discern that spirit entering climate science. The 'Resources' tab of SRCCL's web page in particular has an exemplary array of videos, explainers and presentations for educational use.[16] Without such sensitivity to public understanding, the science and policy that follows would become aloof. It would fail to win the political traction so vitally needed to be implemented by the public sector.

The shock to society and the world's economy brought on by COVID-19 may well transform the politics of green new deals. When the UK government, along with Italy as the co-host, announced the postponement of COP 26 on 1 April 2020, the BBC reported Adair Turner, a Senior Fellow at the Institute for New Economic Thinking, as saying:

> The pandemic will also reorder to an extent the priorities for COP26, as alongside the UN climate process countries will be devising stimulus packages for economies hard-hit by the crisis. With low-carbon stimulus as a new priority for COP26, it should be seen as an opportunity

to rebuild economies hit by coronavirus in ways that are healthier, more resilient to future shocks and fairer to a wider range of people.[17]

Otherwise known as Baron Turner of Ecchinswell, Turner is a former director general of the Confederation of British Industry and a former chair of the UK's Financial Services Authority. Who might have expected such a statement from such a source? One can but hope that a legacy of the pandemic might be the forging of a new economic consensus around what needs to be done about climate change, and that green new deals can no longer be boxed in at just one end of the political spectrum.

Private Sector: Corporate Engagement and Responsibility

Inevitably there are tensions when the voluntary sector tries to work with the public sector. How much more so, then, with the private sector? Given its central position in most economies of today, I reject knee-jerk cynicism about 'the greening' of business and industry. It is easy to point to examples of 'greenwash' in marketing, but, irrespective of our thoughts about how an enterprise might be owned, it would be regretful to overlook some very real achievements that are being made in response to the green imperative. One example is 'dematerialisation', where more is made with less embodied energy and fewer virgin natural resources. Another is the 'circular economy', where waste from one process becomes the raw material for another, so as to make continual use of resources. A third is the principle of 'shared value', where measures like profit sharing, worker representation and beneficial engagement with local communities all point towards a 'triple bottom line' that is social and environmental, as well as financial.

Between 1991 and 2004, I campaigned to stop the highest and most majestic mountain in the National Scenic Area of South Harris in the Outer Hebrides being turned into a 'superquarry'. It would have been the biggest roadstone quarry in the world. The plan had been conceived by a British company, Redland, but our long-drawn-out, faltering success in stalling their progress helped to put a dent in their share price. In 1997, this caused them to succumb to a predatory takeover by the Paris-based multinational, Lafarge, the biggest cement and quarrying company in the world. Through serendipitous circumstances that have been described elsewhere, it fell to me, working closely with local community leaders, Friends of the Earth Scotland and WWF International, as well as other partners under the heading of Scottish Environment LINK, to negotiate what we called a 'dignified exit strategy' from the Redland scheme that they had inherited.[18] Put in less diplomatic terms, we stopped the quarry.[19]

However, in the course of the battle a curious mutual respect built up which led to me being invited onto their Sustainability Stakeholder Panel. 'What? Your greenwash committee!' I said. However, I consulted with the other key individuals and organisations that had been co-campaigners. All said the same. 'Do it. We need to start engaging with industry. But only accept expenses and don't take their money.'[20] On that basis, I served on the panel for the best part of a decade until 2013, by which time I had done as much as I could sustain on an unpaid basis. Other panel members included Jean-Paul Jeanrenaud who directed WWF's One Planet Leaders programme for the greening of industry, a representative of the United Nations Environment Programme, and the heads of two of the major European building workers' trades unions. Working closely with them provided fascinating insights. Lafarge back then knew that their industry needed to change, in the same way as, today, the asset managers BlackRock warn that the energy business is about to undergo 'a fundamental reshaping of finance', with fossil fuel investments that had enjoyed cash-cow status now needing to have 'climate risk' factored in.[21]

To stay ahead of the game, Lafarge responded to robust push and gentle pull from the panel. It shifted its mission statement to 'sustainable building solutions'. This it backed up with extensive research and development into green building technologies, resulting in new product lines. As an example of the circular economy, at Lippendorf in the former East Germany the local power station burnt filthy lignite coal that required the flue gases to be scrubbed to prevent them causing acid rain. The waste product was calcium sulphate, or gypsum, which happens to be the raw material of plasterboard. Lafarge put up a factory next door. It was astonishing, on a site visit of the panel in 2009, to see the gypsum sludge settled out, transported the short distance by conveyor belt and, after suitable treatment, squeezed between two huge rolls of cardboard, dried with warm air and finally chopped up by giant guillotines to yield the finished product ready for shipment. The conveyor belt along which the drying process was carried out was so long that the factory workers rode up and down on bicycles to make their inspections.

Cement-making accounts for 4–8 per cent of world CO_2 emissions, depending on how it is counted. By introducing advanced kiln technology, Lafarge succeeded within about ten years in cutting the carbon footprint per ton of production by one-third, as well as phasing in biodiversity and restoration plans for all of their quarries worldwide and many other measures.[22] In 2018, a 'transition pathway' study from the Grantham Institute at the London School of Economics on greenhouse gas emissions from the world's biggest cement manufacturers, reported: 'In 2020, seven out of the 10 companies with performance data will not be aligned with any of the Paris Agreement benchmarks. Only Ambuja Cements and Lafarge Holcim will be in alignment . . . with the 2 Degrees benchmark, but not the Below 2 Degrees benchmark.'[23]

The people that I knew in Lafarge – now merged with its former Swiss rival Holcim – have mostly moved on. But if anybody is still listening out there, my comment from the posthumous panel would be this: Compared with rivals, you've got the edge on CO_2. Push on towards the 1.5°C target, and use your private sector clout to help public sector regulation to raise the bar of the competitive level-playing field. Help to legislate the dirty players out of business. The pressure cannot slacken until every cement plant in the world is fitted with carbon capture and storage.

The Vernacular Sector: Capitalism and Its Alternatives

Once, when I was giving a guest lecture at Exeter University, some of the students took the view that environmentalists should have nothing to do with the corporations. 'We should be working for their downfall.'

Tongue in cheek, but not entirely so, I said: 'Would any of you who are wearing a corporate product, please remove it now.'

No nudity in class! I eased onto a less stark tack.

'In that case, would any of you who are *not* wearing a corporate product, please stand up.'

One person rose, and there he stood! Resplendent, if a tad eccentric, in a long woollen gown that had been hand spun and knitted, he explained with an impish smile, from his own small-holding's Jacob sheep. It was Timothy Gorringe, the professor of divinity, who writes in one of his books: 'To the objection that the proposals for an alternative economy are utopian I reply, first, that nothing is so wildly utopian as to try and build a sustainable world on the basis of greed and competition.'[24]

But whose greed, and whose competition? My point was that we're all complicit. Partly it is out of lack of choice. But then, what about all those designer fashion labels? What about the driving of a competitive ethos if we shop around to get the best deal rather than the fair price from some wholesome trader? What if we have money in the bank that might 'earn' interest, or pension rights invested in a stock exchange portfolio? The cycling and recycling, the lightbulbs and the finding ways of living more with less? All these have a Zen of personal integrity to them, and especially so (with apologies to the lyricist Tim Rice), a Jacob's Amazing Piebald Dreamcoat. But after that it starts to get more compli-cated. If we can't entirely externalise 'capitalism' or 'the system' as something imposed by an elite from out there, what's really going on?

Trade relies on comparative advantage. Both you and I might be capable of growing apples and oranges. Even in Scotland, I could grow oranges of a sort under glass. But because I can grow apples so much more easily, and you in Spain can grow oranges so much more easily, we'll both produce more fruit in total if we each 'do what we do do well.' That's comparative advantage in a nutshell. However, a Spanish amateur economist might quibble: 'But we don't need you. We grow apples in Spain too, you know!' To which my reply might be: 'Ah yes, but if you do what you do well, and I do it my way, then won't we both be rich enough to chip in for a wee dram too?' At which, by now in song, my newfound Spanish friend would reply: 'We'll raise a glass to that!'

Which is all very well and good. But there's some bum notes. It only works because of cheap fossil fuels that enable ships to shunt apples one way and backload oranges the other. Our economies are both less resilient, because the Spaniard is now dependent on me for apples, and I've developed a taste for luxury fruit imported from afar. Gone are the days when, as an old man once said to me on Lewis, 'An orange was for Christmas.' Such is classical economics. But when people speak of 'capitalism', the whole process escalates. Instead of being there to lubricate exchange, money becomes an end in its own right. Such is the 'advanced' or the turbo-boosted neoliberal ideology of economists like Friedrich Hayek and Milton Friedman that gained a political foothold with the election of Margaret Thatcher in 1979, and Ronald Reagan in the USA the following year. Here, the 'liberalism' in question is not a liberal attitude towards minorities, human rights or the poor. Rather, it is about:

Economic individualism, built on competitive self-interest, where the ownership of property and other forms of capital equates to liberty;

Free trade, to optimise comparative advantages in production, liberating trade from protectionism, regulation and tariffs;

Free flows of capital, so finance can both be accumulated and shift across borders, putting control of the economy beyond the easy reach of governments;

Deunionisation, which subjects workers to wide-ranging market forces, exacerbating inequality, instrumentally treating them as 'operatives' and replacing personhood with 'human resources'; and

Deregulation, to 'get government out of business' and in so doing, to undermine the tax base and the social confidence that keeps governments in business – and with it, the democratic process.

It's one thing to characterise advanced capitalism like that, but quite another to see where it's coming from. The fact is that in democracies, people have voted for political parties that advance these policies. While some might say that the rich farmers have persuaded the turkeys to vote for Christmas, the turkeys might consider such a characterisation to be an insult to their freedom and intelligence. They might rather like what they voted for, which is why they have carried on doing so, and in 2019, at least in England, why they so decisively chose Boris Johnson and his flagship Brexit.

My view – and my bias is towards the psychological and spiritual – therefore sees neoliberalism more widely than party politics and in ways that are more disturbing. Most of us don't realise how tiny individual actions aggregate into systems with emergent properties. We think that money has some kind of objective reality, that accountancy is black and white, but we don't realise that what we're really working with are collective psychological flows of confidence. Gold only has value because we value it. In her poem, 'We Alone Can Devalue Gold', Alice Walker says: '. . . if your chain is gold / so much the worse for you.'[25] Partly then, the trap is thrust upon us, especially if we have no options. But partly, we imperceptibly think our ways into being trapped. The system then takes on characteristics to which most of us have unwittingly contributed.

In his review of *This Changes Everything*, Naomi Klein's book on climate change and capitalism, the philosopher John Gray says that he is unconvinced that global elites are in charge of the world as much as the rest of us tend to think they are. They too are victims of their circumstances, and while that might sound like letting them off the hook, it might just have the virtue of humanising a force in need of being tamed. Gray points out that it is not just market economies that have environmental problems. The command economies of the former Soviet Union and Mao's China also had them. Without elaborating on it further, he suggests that not to face the possibility that deeper factors might be at play, 'shrinks from facing the true scale of the problem'.[26]

What might be that true scale? I notice that the list of attributes above are all profoundly materialistic. There is no hint of deeper values, no altruistic spirit or love of life. Perhaps part of what we have to do is to look into our own internalised capitalism, those Persian two devils, the one we know we've got and the one whose properties are emergent. There's also something here about the psychology, or if we might say, the spirituality of grace. A graceless spirit of utility and exploitation may put cheap products in the shops. It may give us lots of stuff with which to mask the boredom of a shallow life and to feed consumerist addiction. But if not off-set by gratitude, it will block the greater cycles of life's grace, the flows of providence or 'provide-ence'. This causes haemorrhage of empathy. Our consciousness begins to close down. Soul is replaced by soullessness, *satya* by *asatya*.[27] What's left turns to toxin and, not least, as greenhouse gases. We're left with an underlying loneliness and low-level depression. We've come adrift from what the poet Adrienne Rich speaks of as '. . . the true nature of poetry. The drive to connect. The dream of a common language.'[28]

To pay a proper price for something is to ensure that right relationships, both with producers and with nature, have been followed through the value chain. When we say, 'that's too expensive,' we must be on guard lest what we might really be saying is that the cost of social justice and environmental sustainability is too high. Unless we're poor and genuinely can't afford it, mean-spiritedness or lack of care and attention is another way of saying that we'd rather be stuck with – capitalism! Then, perhaps, we try to compensate through what Paulo Freire calls the 'false generosity' of charity. False, because it substitutes for justice.[29] However, if the whole system flipped state, if most of us became sufficiently awake to vote for things like living wages, high animal welfare, low emissions and strong environmental protections, then economies of scale would make the doing of the right thing so much easier. We move in part together with the society around us. If we all start moving, right-on alternatives would become the new norm, accessible and inclusive. The poor would have the most to gain because material poverty is a lack of choice, a vulnerability to exploitation.

In the maritime community that raised me, we were taught you don't abandon ship until you've got the lifeboat in the sea. In these matters, that's why I'm not a revolutionary but an evolutionary. As compromised people in a compromising world, each of us navigates from where we're at with one eye to the ground, but the other to the stars. That's also why I don't belittle middle-way paths like corporate social responsibility. But what might the dreamed-of destination of a world of right relationships look like? By what signs in the sky might shepherds watch for its emergence? For me, this is the importance of movements like Fair Trade, community initiatives, the gift economy, organic agriculture, and all that revolves around what I think of as the *vernacular economy*. I use that term to bring a particular focus to bear on the range of what are variously called new, ecological, alternative or autonomous economic theories.[30]

The Croatian-Austrian thinker Ivan Illich wrote of 'vernacular values' as those that, like our mother tongue, we learn and practice informally in our families, communities and the bioregions where we live. He pointed out that 'vernacular' is derived from an Indo-European stem that implies 'rootedness' and 'abode'. It represents, he said, 'autonomous, non-market related actions through which people satisfy everyday needs'.[31] Curiously, the seminal essay called 'Vernacular Values' in which he explores all this, was first published in 1980 by Stewart Brand in *CoEvolution Quarterly* – the magazine that also ran R. Crumb's cartoons on how we've got to where we're at.

If we choose, metaphorically, to 'devalue gold' or, at least, to keep our involvement with it within discerned confines of the golden mean of proportionality – another of Illich's themes[32] – then we start to shift our own internalised capitalism. Then we start both to confer and to receive blessing. I accept that some people will disagree with that. Some will say that it puts too much burden on the individual, and that what we need is a complete system crash and then to press the reset button. Perhaps. But I look at the trauma set loose by the 'reset' of war, and opt instead, as just said, for the evolutionary approach. Although not obviously caused by climate change, shock events like the coronavirus pandemic will nudge us on our way if we're ready to respond wisely. In so doing, what might be our guiding star? We'll come shortly to modern Scottish land reform as one such pattern and example of the vernacular rejuvenated. For now, let it suffice for me to point towards a passage that I love in the writings of Alfredo López de Romaña, a Peruvian 'barefoot' economist:

> Life must be the center of the new social order. Life is the common denominator of the various 'social movements' that have been growing in opposition to the megamachine's expansion . . . It is Life and not things, machines or power that must guide social organization. Things and technics are relevant only insofar as they enhance life. And it is the vitality of its people, the moral elegance of its culture, and the poetry of its economy that make a nation great – not its power.[33]

There we glimpse the dream, the quadruple bottom line. But lest we just dream on, let us attend first to the scale of the sea of nightmares that climate change whips up. Even to attempt to get a lifeboat in the water, we need to know how fast the ship is going down, and what and whether anything might buy us time.

Limits to Growth and the Physics–Politics Gap

In 1972 a group of 'futures thinkers' called The Club of Rome produced a pioneering computer model of the world's population, resources and economy, resulting in a study called *The Limits to Growth*. Hugely influential, it nevertheless fell under attack from several quarters: from business, because it challenged the ideology of growth; from some religious groups, because it put population on the table; and from the left, because it seemed to question whether affluence could be a dream for all.[34] The business professor Julian Simon, a renowned critic of new economics, said that it failed to allow that what constitutes a 'resource' changes with time.[35] Cave dwellers in the Stone Age probably worried about running out of flint, and true enough, on the model's basis, several of the world's critical natural resources should by now have run out.

Nearly half a century later, computer modelling has advanced and technological developments have clinched Simon's point. What's changed is that today the need for limits is more on greenhouse gas emissions. A country like Scotland may have achieved a 47 per cent drop in its emissions that are attributable to electricity generation between 1990 and 2017, but that's with plentiful renewables resources. The higher-hanging fruit of emissions from agriculture, industry, transport and heating are going to be very much harder to reach. In consequence, the 2019 report of the Committee on Climate Change warned the Scottish Parliament that its 2030 target to reduce emissions by 75 per cent, 'will be extremely challenging to meet'.[36]

Even more to the point, and on a global scale, is the UN Environment Programme's *Emissions Gap Report*. It states that during 2018, greenhouse gas emissions leapt by 2 per cent, and that they'd averaged an increase of 1.5 per cent per year over the past decade. As it bluntly concludes: 'The summary findings are bleak.' When everything is factored in – emissions from fossil fuels, industrial processes, agriculture and other land use, and other greenhouse gases expressed as their CO_2 equivalents – the total CO_2eq figure for 2018 was a staggering 55.3 billion tons. Given that by 2030 these need to be 32 billion tons lower to attain the 1.5°C target, the report states also, in a masterstroke of gnomic understatement: 'The emissions gap is large.'[37]

A major study that frames the emissions gap picture for the UK is *Zero Carbon Britain*, released in 2019 by the Centre for Alternative Technology in Wales. It looks at how it would be technically possible for Britain to decarbonise, at what it calls 'the physics' of the possible, but it quickly runs up against what it calls, 'the physics-politics gap'. Namely,

> if we analyse these physical requirements and work out a physically credible plan based on our scientific knowledge of the situation, we find it does not fit comfortably into the frame of normal politics and economics . . . In fact, a huge gulf between what is physically demanded by science and what is seen as politically possible is revealed.[38]

On the one hand we have the optimism of the ecomodernists: that technofix will one day fix us. On the other hand, what's happening to the earth is already tracking high on some measures, relative to IPCC mid-range forecasts. For example, higher than expected rates of Antarctic ice loss threaten to add an additional 10–15 cm to the rise in global sea level by 2100.[39] While the coronavirus will put a dent in emissions (but not in the ever-accumulating level of concentrations), as happened after the banking crisis of 2008, the global economy will probably recover. If so, where would such a picture leave us?

There is a passage in T. S. Eliot's *Four Quartets*. The underground train with its contingent of merchant bankers, distinguished civil servants, the chairmen of many committees, and the petty contractors – the you and me, we might imagine, on zero-hour contracts – goes into the dark, dark, dark, 'the vacant interstellar spaces, the vacant into the vacant', and stops for too long between stations. The conversation rises, slips away again to silence, 'and you see behind every face the mental emptiness deepen' and only the 'growing terror of nothing to think about'.[40]

What can one say? What can this dear world do, caught up in the emergent properties of its own predicament? The locomotive that pulls human life seems very broken down inside the tunnel. Perhaps a starting point might be to revisit how the climate change debate is framed. Perhaps we have some blind spots, some angles that invite fresh exploration, and maybe even, modest openings of the way.

Population, Consumption and Sustainable Development

When discussing SRCCL – the IPCC's special report on climate change and the land – we touched upon the twin drivers of population and consumption. Recognising that the carbon embodied in consumption will be subject to moderation by technology, we can nonetheless derive a formula, in the most simple back-of-an-envelope terms, by which:

Population x Consumption = Greenhouse Gas Emissions

A letter in *Nature* suggests that to stabilise the world, we need to return to the per-capita emissions level of 1955, the year in which I was born.[41] Taking that at face value for argument's sake, we can see just how recent is the full scale of the problem. Since 1955, world population has nearly trebled. CO_2 emissions from consumption have risen six-fold. Consumption has therefore twice outpaced population. Some of this is lifting people out of poverty. The rest can be seen in the hotel car parks, berthed in the luxury marinas and in the mansions on the hill where, as John Lennon had it, 'there's room at the top / they are telling you still.' There is no question. Tackling the carbon embodied in material consumption is of greater magnitude and consequence than population as an issue.

However, even to mention 'population' is sensitive. It is a wolf whistle to the hard right and spells racism to the left, linked in many people's minds to 'population control' and, in some countries, to traumatic memories of forced contraception, compulsory abortion, eugenic sterilisation programmes and male control over female bodies. Even in Britain in 2020, an advisor to the prime minister Boris Johnson was suggesting that 'one way to get around the problems of unplanned pregnancies, creating a permanent underclass, would be to legally enforce universal uptake of long-term contraception at the onset of puberty'.[42] He subsequently resigned, but such attitudes signify the dichotomy between some of those who presume to control others and those whose lives might be controlled by them. It undermines the dignity of all that is *swaraj*, that is, of freedom and dignity, and if extended to the politics of climate change it panders to the notion that the problem is not the consumerism of the rich but one of 'too many' Indians, Chinese, Muslims, or whoever else can be 'othered' as the scapegoat of the day.

In Chapter 3 we saw that IPCC makes only passing mention to both population and consumption. Nevertheless, assumptions are necessarily built into its reports and models, just as assumptions about a neoliberal world economy are built in as being the way things are. For example, SRCCL states that the Paris Agreement formulates its goals for limiting global warming 'while factoring in the need . . . to accommodate a growing human population'. However, it contains only scattered mentions of such terms as 'over-consumption'.

SR1.5 goes a little further. Under a subheading on socio-economic drivers in 1.5°C pathways, Chapter 2 acknowledges that world population by 2100 could range between 6.9 and 12.6 billion. Significantly, it states that an important factor in determining these differences is 'future female educational attainment, with higher attainment leading to lower fertility rates and therefore decreased population growth up to a level of 1 billion people by 2050'. In other words, women's education alone could make a difference of something like 10 per cent, and while the effects of this on greenhouse gas emissions would not be one-to-one, because affluence and life expectancy would probably rise in tandem, the general point remains that a lower global population would mean fewer people to share any carbon budget factored in to future climate targets.

Earlier, we saw Mike Hulme 'deny that there is warrant to collapse the possibilities of the future to human extinction'. We might now ask if a consequence of alarmism's theft of time, of its shrinking of the time horizons, is that it narrows the system boundaries that could allow population and consumption changes to alter the framing of the debate.

Population, Fertility and Women's Equality

We will look at the drivers of consumerism in the next chapter. But for now, we need to ask if it is possible to approach the population question without falling, or being dragged, into the bygone traps of 'population control'? I think that it is. I think so for the reason, as is nodded to in SR1.5, that it fundamentally ties in with the free choice and liberation of women in particular, and of people in general. How so? Because multiple studies suggest that the key factors that correlate with falling fertility rates are such desirable qualities, as:

Respect
Equality for women
Education for women
Voting rights for women
Economic opportunity for women
Urban or well-connected rural living
Family welfare and security in old age
Good perinatal care and low infant mortality
Effective family planning services and late family start
Resolving patriarchal domination and violence in society

In other words, dishing out the pill, or condoms, or tuition in 'natural family planning', is not the front line of what is needed. It can even be a divisive distraction from the main task at hand. What we need to look at instead is a wholesale programme for a different world, one that is most unlikely to be appropriated by authoritarian political or religious factions. While gender equality is not a solution on its own for climate change, in the medium to long term it does have a huge contribution to make towards drawing human impact on the earth down within planetary carrying capacity. What's more, women's equality benefits men too, for we men have also been oppressed by our own patriarchal structures, particularly where militarised.

Neither is our choice limited to either technofix or the hair-shirt-in-a-cave life. I believe that we need a middle way between these two, a wise affluence of simple but dignified sufficiency in our consumption of material things. War has legs and peace must travel too. It is not unimportant that people should still mingle, trade and even marry across cultures and geographical regions. To live as a One World village therefore invites a modest but not excessive level of connection and exchange. Just how we do that will depend on the balances between population, consumption, technologies and emissions.

The problems that we face must not be cast in terms that undermine the wherewithal for the said dignified sufficiency of living. That is a demand that we must try to meet. The problem is with the insatiable excess of the never-satisfied consumer and the commercial forces that push insatiability. If, from lack of vision, ingenuity and will, we undermine wellbeing, the politics will never work. However, the good news is that the framework to achieve what needs to be attempted is in place. It is already highlighted by the IPCC. Such is the UN's 2030 Agenda for Sustainable Development. For those who might feel reassured by making the medical aspect of this explicit with respect to population, Goal 3 – Good Health and Well-being – aims by 2030 to: '. . . ensure universal access to sexual and reproductive health-care services, including for family planning, information and education, and the integration of reproductive health into national strategies and programmes'.

Let's round this off by taking a look at just how dramatically fertility rates can fall once the right social structures start dropping into place. Bearing in mind that, depending on a country's child mortality rates, the fertility rate needs to average about 2.1 children per woman to achieve population replacement, the estimated national fertility figures as of 2020 are, for the UK 1.9, India is down to 2.3 (and just 1.8 in the progressive state of Kerala), China is at 1.7 notwithstanding having abolished its one-child policy in 2016, and Malaysia, which is predominantly a Muslim country, stands at 2.0. What about culturally Roman Catholic countries, so often the whipping-boy for the 'population control' lobby? Well, no longer so. Ireland is now down to 2.0. Brazil stands at a remarkable 1.7, Italy is even more surprising at 1.5 and Portugal is the third lowest in the world, at an eerily depleted 1.2.[43]

Moldova is the same, this being a landlocked agricultural republic that is sandwiched between Romania and Ukraine. There, 97 per cent of the 3.5 million population identify as being mainly Orthodox Christian.[44] Neither can it be said that low-population countries are necessarily rich. Moldova is one of the poorest nations in Europe, a former Soviet satellite, not a member of the European Union, with a per capita GDP of only some $3,000. It doesn't even have particularly good family-planning services. The main driver of its worryingly low fertility rate is that in the majority ethnic group, women are free to choose careers. They start their families late, if at all, and many choose to have only one child.[45]

In painful contrast, countries that have high fertility rates are typically those where women are kept under the thumb of the patriarchy, with low status and few freedoms. Usually these countries are war-torn, intergenerationally traumatised and impoverished. Niger currently tops the league at 7.2 births per woman of childbearing age, followed by Somalia at 6.1. When I looked at the figures in 2009, the fertility rate of Afghanistan was right at the top, at 7.2. Today, because the position of women has improved as the Taliban's control shifted, it has nearly halved to 4.4.

I stress and cannot stress enough that the population element in how climate change is framed is only one factor. Personal consumption is the greater one, and then there is the interplay of technology in decarbonising energy and manufactured goods. We have seen, too, how policy measures can come into play through the public and private sectors of the economy. All of these interact, but on the population question and the religious concerns that can sometimes stop it being aired: quality of life can count for more than quantity. While the Hebrew scriptures may have said to go forth and multiply and fill the earth, they never said to overfill it. The time has come for the environmental movement to stop sidestepping these twin elephants of population and consumption that fill the room. Like with nuclear power, there are no easy answers; but I believe we do require ongoing fresh evaluation in the light of ever-changing pressing circumstances.

Untangling the String, Clearing the Desk

There is one other thing. Falling birth rates create their own problems, not least in how to support the care needs of aging populations. Some countries – Finland, Estonia, France and Russia amongst them – have introduced incentives, particularly to have a third child.[46] At the same time, for all the uncertainty around climate change as a current driver of refugee flows, people's need to find new homes in relatively habitable parts of the world is likely to be a major emergent issue.

In Scotland, there are traditionally two sacred duties to the guest: hospitality for the short term, and fostership (or adoption) for permanence.[47] As a proverb has it, 'The bonds of milk are stronger than the bonds of blood.' Nurture counts for more than mere blood lineage. What if nations were to dig into their treasuries of poetry, song, literature, mythology and spirituality, and draw out oft-forgotten material from which to strengthen multicultural values?[48] After all, as Christian groups may or may not be aware, Jesus was only able to do his thing as prophesied as a member of the House of David because Joseph adopted him into it. Furthermore, the holy family had to flee a reign of terror as refugees to Egypt.

Such a story – whether we take it literally or at a deeper level of metaphorical meaning – is a challenge to complacency. It might raise the question: What if we were to take climate change not just as a threat, but as an opportunity to deepen our humanity? What if we, who might be of the countries that historically have done the most to cause climate change, were to find it in ourselves to welcome climate refugees? And to do so not just as cheap labour, but as friends and fellow citizens? If we treat others like ourselves, they'll come into the family. Who knows what gifts the welcomed stranger bears.

In the early 1990s, courtesy of his daughter Jane King who was then a colleague at the Centre for Human Ecology in Edinburgh University, I had the good fortune to spend a couple of hours with her father, Alexander King. Then in his eighties, King senior had been Director-General for Scientific Affairs at the Paris-based Organisation for Economic Co-operation and Development, and had also co-founded The Club of Rome, the think tank that commissioned the 1972 *Limits to Growth* study. I leapt at the opportunity to ask what advice he'd offer to a young scholar–activist. His reply was to remember that the social and ecological systems of the world are like a tangled ball of string. If you pull on any one loop, you'll find that it's connected through to all the others. And he also said, always keep a space clear on your desk for the unexpected.

In the next chapter, I want to use that space to entertain the unexpected, to enter into a very different mode of discourse. I'd like to try to unravel some of the most deeply tangled loops that contribute to driving consumerism in our psyches, and to look at land reform as one example of a way ahead that can change the people and environment relationship. I realise the inadequacy of such a narrow focus, but I am also mindful of the power of inspiring stories and the way that all of us must dig from where we stand. As with making the proverbial stone soup, if we can all add just one ingredient, we can end up with a rich broth round the hearth.

I will do this through a framework of four Cs: *clearance* or colonisation of our relationship with the land and one another; *collapse* or trauma that historically resulted and which arguably lingers on across the generations; *consumption* as being not just for the satisfaction of fundamental needs, but as consumerism that fills an inner emptiness; and the rekindling of *community*, as a means of restoring flows of life. That, for the survival of being, both human and nonhuman on this planet. These might seem like an odd assemblage, but they offer me a framework to dive down very deep into that inundating murky river of the dream mentioned earlier. What follows will be neither be definitive nor exclusive of other points of view, but I do promise, it will contain surprises.

Much of what I share from here on will be impressionistic, a case of showing and not just telling. Leave this book at this point if you only need the material of the earlier chapters. Otherwise, come with me if you wish. We'll rejoin people on the land along the Hebridean scenic route.

THE SURVIVAL OF BEING

Back at the offices of the South Pacific Appropriate Technology Foundation in Papua New Guinea, Andrew Kauleni of Tuvalu sea level fame had pinned some lines of poetry to his wall. They'd been written by his friend and one of the Foundation's trustees, the human rights lawyer Sir Bernard Narokobi, who was for a time the country's minister for justice and later, its high commissioner to New Zealand.

When Bernard died in 2010, obituaries hailed him as an icon not just of the nation, but of the entire region. He personified a philosophy of life that he wrote of as 'the Melanesian Way', the indigenous wisdom of the peoples of the south-west Pacific islands.[1] As I remember it, his lines went:

Welcome to the University.
The ancient, timeless, eternal
University of Melanesia.
The village,
where courses are offered in living.

Let's then open the semester. In so doing, we will ground ourselves in people and place, and move progressively from the outer to the inner life. In this way, we might see in a more visceral way than abstract discussion can permit the qualities that might help to face the come-what-may in the come-to-pass of climate change.

Where Courses are Offered in Living

The air over Loch Leurbost rang with the wheel of gulls and the trill of waders that had come to feed on crabs, shrimps and little fish exposed upon the falling tide. Our island village hosts with their Papuan guests – two cultures, half a world apart – had moved on now from the ancient forest of the exposed pine stumps where we reflected on the changes wrought in part by climate change.

We came to an inlet that was filled with village fishing boats, some still tied with hawsers to their mooring posts. You'd think the knots were made just yesterday, but every hull was rotting, holed and sunken. Old engines languished, Listers they would usually be, seized and rusting on their bed frames. Abandoned anchors poked out from the tangled grass that granted such nobility of purpose no self-respecting depth of grip.

I thought about my time in Papua New Guinea some forty years before. I'd read an autobiography by Michael Somare, the newly independent nation's first prime minister, who came from the Sepik River area that neighbours Indonesian Papua – the same as Andrew Kauleni and Bernard Narokobi. The tensions in Irian Jaya, as the western half of New Guinea was then called, were never far away from discussion in their company. Somare told the story of how his father and grandfather had taught the way of *sana*, the art of being both a warrior and peacemaker. They'd taught that every clan had its own magic, 'and ours is the magic of peace'. He said that when people came to have a tribal fight, that was OK, they could have their fight. But first they had to sit down on the beach and have a feast and everyone would talk. If they still felt like fighting afterwards, that was OK too, but: 'We believe that after eating, their minds will be changed.' And he taught that when you see a person trying to pull their canoe up the beach, you should go and help them, invite them in, because '*sana* invites people'. And you know what? 'People will always remember the man who helped them to pull up their canoe.'[2]

One of our guests, Wahyudin, stood there at the tapering neck of the sea loch, shaking his head in amazement. He'd grown up on another Indonesian island, Sulawesi, which was famous for its seafarers and has a population density of 300 per square mile. Both Lewis and Papua have densities of nearer twenty-five.

'Where I come from,' he said, with our colleagues Adrian and Maria doing the translating, 'somebody would come in the night and take those for scrap.'

'Not here!' quipped Evelyn, who lives nearby and works in Gaelic cultural television. 'Everyone would know who did it.'

The men who last tied up the Leurbost fleet of boats were all contemporaries of ours. Rusty knows which one belonged to whom. Around the early 1970s, new technologies, in lockstep with new greed, swept through our sea lochs, the same as is happening across the Pacific today. Within a couple of years, our fisheries close inshore had disappeared, as if a light switch had been tripped. They say about the working of the land that 'a lot went out with the horse'. The same was true of village boats. Another mooring line had snapped that once connected the community. One less living context in which to make our *sana* and to cultivate a shared resilience.

We move on from the bay of the fishing boats, on over the long waves of *feannagan* that bleed their soil into the sea, a few crumbs at a time on each new storm's high tide. We cross a wall that runs into the sea, and that's the marker. That was the place deemed far enough away from household drainage before they put the village sewage system in during the 1960s. From this point on, Alex George had told the rest of us, the shellfish were considered fit for human consumption. He was the oldest in our year from our end of the village and knew about things like that. My dad, the parish doctor, agreed with his assessment. In ecology, you can take the precautionary principle to its extremes and not eat. Or you can balance up the risks. Today they've got a name for it all – 'traditional ecological knowledge' – and, maybe mindful of his role, it had been Alex George who'd said to me months earlier to go and speak to Rusty, and to bring the Papuans to this place. It wasn't like they'd parachuted in, they'd had an invitation.

But why come so far, 9,000 miles, and in the face of climate change? Why not take them just to neighbouring Papua New Guinea, where I still have the connections? Suffice to say, and to leave it there, the border is not open. I'd met Maria Latumahina at Schumacher College while running a course on land reform. A native of West Papua Province, she was working for the British Embassy in Jakarta. She and her colleague, Adrian Wells of Anglo-Tamil-Malaysian background, had procured funding from British and Indonesian government sources for training civil servants in climate change awareness.[3] One outcome of earlier study tours was provincial legislation being passed to strengthen indigenous people's rights over their forests, protecting 90 per cent of them from development as oil palm and other plantations. Another outcome, was that Alex Rumaseb, the most senior figure behind these delegations, published a series of books locally on small-scale 'crofting' land use, collaborative leadership and liberation theology for bottom-up empowerment.[4] The outcome was a widening of the horizons of possibility, not just for the Papuan guests but also for the host communities. To learn to be One World, it helps to see One World.

Anybody who knows the history of that part of the earth will know how significant such widening of horizons is. Early Dutch colonisation of Indonesia was a brutal business. Slavery in the Dutch East Indies, as the scattered colony was known, continued until 1860, with people ruthlessly cleared from their ancestral lands and set to work elsewhere as plantation labourers. Two days after the Japanese surrendered in the Pacific in 1945, independence was proclaimed. For a nation whose main common denominator had been a shared colonial master, this soured into the Sukarno dictatorship. The American-backed communist purge of 1965–6 left half a million to a million dead. A whole generation of artists, writers and intellectuals were wiped out. Even though Indonesia has since become a democracy, the trauma festers on as a suppurating scab, its harrowing cruelty laid bare in the 2013 documentary film, *The Act of Killing*. Democracy alone – impressive though the achievements have been in recent years – cannot come up with instant fixes for such a violent legacy.

The Papuans had therefore been clear that they wanted to build an understanding of what had happened to them and how they might most wisely move forwards. Ever since 2012, when we took the first such delegation to the Isles of Eigg and Skye, we had worked on their agenda by loosely linking these principles to climate change. We had looked on it particularly through the lens of drivers of consumerism, this being such a prominent symptom of the breakdown of relationships between communities and the earth. At first, I'd wondered if the links were maybe too tenuous, even though on Eigg, where I'd played a part in the pioneering 1990s land reform campaign, 95 per cent of their electricity now comes from their own solar, wind and hydro power.[5] But the more that we dug into the wider issues of the world, the more we connected the local to the global and saw the patterns of a common human cause.

So here we'd come. This time and with this delegation, to Loch Leurbost. And unbeknown to them, stretching southwards from the boundary marked by the river that we'd just crossed, lay a social history that went deeper, and with a greater contemporary global relevance, than most could ever have imagined from a place so emptied out of present-day habitation.

Clearance – What We've Lost

We reach some boulders on the shore beyond the wall. Evelyn takes the lid off her yellow pot and decants the kindling. Nobody says anything. Those amongst us who know what to do just begin in harmony of instinct to lift aside the seaweed tresses and start to fill the pot with gorgeous blue-shelled mussels, each of them a thumb's length and half again its width. The making of community isn't rocket science. 'Only a demanding common task builds community,' said George MacLeod, who founded the Iona Community. This bit isn't even demanding. The Papuans require no telling and join in with whoops of joy. They too know what to do. We wrench the shellfish off the rocks with deft twists of our wrists. They clang into the pot, and all of a sudden, in this *sana* of shared gathering of the food, indigenous comes face to face with indigenous. Canoes are pulled up all along the beach from right across the world.

'This is amazing,' says Kristian Sauyai, a stocky maybe forty-year-old, who spoke a little English. 'This is how we'd welcome visitors in Raja Ampat. We'd take them to the reef and we'd all collect the shellfish for the feast.'

We continue over to the clachan, to the ruined village like a row of knocked-down dominos. The two most senior women, Mama Enggelina and Mama Morin, go on downhill with Catherine Mary, carrying the pot between them. She was young enough to be their daughter. Down they go, on down to a little stream that flows out from the hills, and there they wash our harvest in fresh water. Meanwhile, Evelyn and I arrange the kindling in a ring of tumbled stones. We light a fire just big enough to steam our feast, pouring just a single cup of water into the bottom of the pot.

The others gather round and find their seats and settle in. These grass-covered stone walls are now our ceilidh house, the hearth around which, in the days before TV, folks would tell their stories as if a village university. I've heard it said that 'stories tell us about our past; visions tell us about our future.'[6] We're here to talk about the future, the future as patterned from the past, the future that is in part already carried at unconscious levels in the psyches of us all. What's more, the less a person thinks they have a past, the more a past lies hidden.

Within a half hour, we're passing round the mussels by the handful, relishing their succulent orange flesh. Then Evelyn begins. She dives in at the deep end, her fair hair streaming in the wind, voice steadfast with the confidence that can be found these days in island women of a younger generation, women who are modern and yet have reconnected to the memories and values of those that went before.

'This will be the first time, in a very long time,' she says, pausing to take in the immensity of the occasion, 'that a meal's been served in this house.

'I don't know how far back it was. Today's the first time I've been over here. But this was the home of one of my grannies.'

Attention grips the gathering, '. . . the home of one of her grandmothers,' her words repeat, in Bahasa Indonesian.

'You see,' comes in Catherine Mary, 'in the past, we weren't taught our history.'

'No, we weren't taught it,' echoed Evelyn. And then Rusty – real name Seonaidh Macdonald, though really, I should be listing their patronymics going back several generations so as properly to place them in their lineages – settles in to his historical remit, and begins to tell our visitors why we've brought them here. Perhaps I might relate it, with my own bits added in?

It started around 1600, within a year or two of the formation of the first joint stock companies – the East India Company had a private charter from the English crown to colonise India, and the Dutch East India Company assumed the same control over the island groups of many different cultures that would eventually become Indonesia. At the same time, in order to colonise and privatise the Isle of Lewis, and to gain control of its rich fisheries, the entire island of some 700 square miles was, by force of fire and sword, taken off its clansfolk – mainly the Macleods, Morrisons and the Macaulays – and traded by King James the VI and I to a compliant mainland chief, Mackenzie of Kintail and Seaforth.

Fast forward a couple of centuries, and a subsequent heir to the estate, Colonel Francis Humberston Mackenzie, wished to curry favour with the fledgling British state. Two miles south from where we sit there is a rock, a landing point upon the shore, called *Leac nan Gillean*. It means the Rock of the Young Lads. In 1808, and with the seeming acquiescence of the established church minister who was appointed by Mackenzie under landed power of patronage, a Royal Navy press-gang ship closed in. The community had been assembled inside their church. The doors were locked and, under armed guard, more than thirty of the young men were marched to the rock and forcibly shipped off for military service.[7] Few if any would ever return. Similar events were happening right across the South Pacific and on through into the early twentieth century. The Papuans later reflected that they'd heard stories from their grandparents about their forebears being kidnapped to be porters by the Japanese in the Second World War. Reporting on a Royal Commission in 1885, the premier of Queensland described the practices of the time as being 'as bad as the African slave trade at its very worse', but the government was committed to providing plantation labour.[8] The capture of Pacific native peoples went by the name of 'blackbirding', as if it was a bloodsport. The only difference that would have limited that term's applicability to this part of an emergent United Kingdom, would have been skin colour.

Mackenzie's penchant for press-ganging laid the foundations for what became the Seaforth Highlanders regiment. In recompense, the British state made him governor general of Barbados from 1802 to 1806. In his slave plantation colony, 4,000 miles from his Hebridean possession, he ran up gambling debts. In a bid to pay them off he leased the vast expanse of Lewis from south of the river that we'd crossed, and gave it over to commercial sheep farmers from the mainland. Demand for wool was high to make trench coats for the Napoleonic wars. The heartless entrepreneurs who took up Mackenzie's leases cleared the ground in a series of ruthless mass evictions. Such 'Highland Clearances', as they're known in Scotland, included the clachan of Croigearraidh ('croy-garry') in the ruins of which we'd had our mussel feast. The stones on which we sat were witness to such history, and the Clearances, just a more polite name for colonisation.[9] Only very recently has a younger generation of scholars felt sufficiently confident to call it so.[10] In England, the same process, taking place further back in history and more obliterated from the folk memory, was known as the Enclosures. The social historian E. P. Thompson therefore called his best-known work *The Making of the English Working Class*. The 'making' was not the natural and inevitable occurrence that most English folks have been led to believe.

Lost settlements like Croigearraidh can be picked out with startling clarity on Google Earth. The land around is hallmarked with long lines of the *feannagan*, and can be seen in nearly every sheltered bay of south-east Lewis that had arable potential and a year-round freshwater supply. The evictions mainly happened in the 1820s, after Humberston Mackenzie's death in 1815. Some families sailed up from further south, and were forced to spend the winter underneath their upturned boats at Crossbost by the mouth of Loch Leurbost. Others walked across the moors to another village, Aird Tong, just north of Stornoway. An eyewitness report from there in 1828 said that their condition: '. . . beggars description. It is worse than anything I saw in Donegal, where I always considered human wretchedness to have reached its very acme . . . The poor people at the new lots there are suffering the greatest hardship, many of them dead, I am told, from disease brought on.'[11]

One of the families that settled in Tong were of the name Smith. They had been cleared in 1826 from Budhanais on the shores of Loch Shell, ten miles south from Leurbost. Rusty thinks his family is connected to that line, but he's still picking up the pieces as the documentation is poor. The grandson of these evictees was a Malcolm Macleod, a crofter who worked what little land he had, as well as being a fisherman and the 'compulsory officer' for the local school. In that capacity, his job was to chase truanting children back into class where they'd be educated, not in their native Gaelic, but in the language of their social betters: of 'civilisation' and its mores. Even into Rusty's days and mine at Leurbost School, an old world map of the British Empire still projected its power from the classroom wall, 'our' parts in imperial red, being those on which 'the sun would never set'. I can remember feeling proud of it. Such was the power of our conditioning.

In 1930, at the age of seventeen, one Mary Anne Macleod, a daughter of Malcolm and his wife Mary, set sail from Glasgow on her own and emigrated to America. She reached New York the day after her eighteenth birthday with just a few tens of dollars in her pocket. There, the newly arrived economic immigrant took up work as a nanny and domestic servant. Being a 'bonnie lassie', she soon met and married a property developer of German descent.[12] They say that a good Scots sermon should have three points and a head. So it was to Rusty's grand finale on that April's day in 2019, as we sat together in those ruins of clearances sparked off by white supremacy from Barbados.[13]

'And one descendant of that family, that used to live in a black house, now lives in a white house.'

Collapse – What Happened to Us?

If it feels like I may have lost my way with climate change along the scenic route, bear with me till I come full circle.

Colonisation is not just the clearance of people's lands. To sustain unequal power relations, it is an ongoing clearance of our minds. A people without a history struggle to know themselves. A person who struggles to know themself will feel insecure in their sense of agency in life. Rootlessness left unresolved can lead to inner disconnects between people and place. The oppressed can then so easily flip to becoming the oppressor, as the Persian second devil sets loose. Narcissism, or pronounced self-centredness, is less a choice than an affliction.

Mary Anne Macleod's fourth son, Donald John Trump, was elected president of the United States in 2016. The book said to capture best the zeitgeist that brought him to power is J. D. Vance's *Hillbilly Elegy*, a *New York Times* bestseller. Himself proudly of 'hillbilly' stock, yet providing the uncomfortable service of being a critical friend to his own culture, Vance's introduction describes what he calls, importantly, the 'spiritual and material poverty' of white Appalachian culture. He says:

> I do not identify with the WASPs of the Northeast. Instead, I identify with the millions of working-class white Americans of Scots-Irish descent . . . [whose] ancestors were day laborers in the Southern slave economy . . . Americans call them hillbillies, rednecks, or white trash. I call them neighbors, friends, and family.
>
> [But] we're more socially isolated than ever, and we pass that isolation down to our children. Our religion has changed – built around churches heavy on emotional rhetoric . . . Our men suffer from a peculiar crisis of masculinity . . . reacting to bad circumstances in the worst way possible . . .
>
> There is a lack of agency here . . . The fact that hillbillies like me are more down about the future than many other groups . . . suggests that *something else is going on*.[14]

His is not an academic study. 'My primary aim,' he says, 'is to tell a true story about what that problem feels like when you were born with it hanging around your neck.'

Scholars of the Scots-American diaspora point out that much of what passes as a 'Scots-Irish' identity is imagined, grounded more in mythic choice than genetic lineage and traditional practices.[15] It is why American clan gatherings can often seem ridiculous to natives back across the Atlantic. But psychologically, the archetypal narratives and imagery that such choice activates are revealing. The search for 'roots', the intense yearning to belong, raises questions as to what became of lost connections to identity. It raises questions of both personal and cultural psychotherapy, that ask: What happened?

In talking about Mary Anne Macleod on the island, we've counted no fewer than seven levels of trauma that affected the community and the specific family background in which she, as President Trump's mother, was raised.

Their 1820s clearance from the land, and that on two maternal sides: her Macaulay line were also cleared from Kirkibost in the west.

Multiple fishing tragedies. Broad Bay at Tong is very exposed. Mary Anne's maternal grandfather drowned in 1868 when his boat capsized in a sudden storm, leaving her one-year-old mother to be raised fatherless.

The First World War, in which a thousand young men (about one-in-six of that age group) died in the trenches or on the Atlantic convoys.

The *Iolaire* Disaster on 1 January 1919, when some 200 naval ratings returning from the war drowned when their ship hit rocks just five miles from her village. Mary Anne was six years old at the time. The entire island was thrown into an unfathomable depth of mourning that was rarely spoken of until its recent centenary.

The Spanish Flu pandemic that followed the war, and to which native islanders had little resistance.

The tuberculosis or 'consumption' epidemic, to which they also had little resistance, and was compounded by poor housing. My father still had patients in the 1960s with its lingering effects.

Mass emigration in the 1920s, especially of those who had not been given land as had been promised after

the war. A thousand left in 1923 alone when Mary
Anne was eleven. On the steamship *Metagama* that
sailed from Stornoway that year, all but twenty of
the 300 on board were men, with an average age of
twenty-two.

Island men in those days married late and girls married early. With
such haemorrhages of her marriageable age group, what choice
would Mary Anne have to pair with? One can see the allure of
emigration when so many of the young men were already dead or
gone. She wrote a letter to a friend that made a passing reference to
'. . . the lonely Isle of Lewis'.[16] One can but speculate, but perhaps,
no wonder. Something about that name so iconic in the island's
oral history, the *Metagama*, seems to say it all: as if beyond the
reach of even radiation. And this is the maternal psychohistory,
not just of President Trump, climate change denier of the world
in chief, but of so many other people; and especially so, as J. D.
Vance's point about 'spiritual and material poverty' nods towards,
from the backdrop of Trump's core voter constituency.

    Little is known about Donald John's childhood. One study,
Trump on the Couch, portrays him as having been 'a fussy, fidgety
baby who would grow into an aggressive, hyperactive child', and
that if Mrs Trump had 'exhibited an especially warm or maternal
character in the raising of her children, it has not been reported'.[17]
Another biography maintains, but without citing a source, that he
had a Hebridean nanny. Whether so or not, his mother's friendship
circle on arriving in New York would likely have majored on the
Hebridean diaspora. There can be an immigrant vigour that comes
from not having to carry a share of extended family responsibilities
back in the village. Once married, she seems to have channelled
her energies into a bon viveur lifestyle. She still spoke Gaelic on
occasions when she'd return to Lewis, but unlike her popular sister
and the remaining family at Tong, the island – while not speaking
of her badly – does not speak of her well. In the island's way, that
speaks for itself.

    In Africa they say it takes a whole village to raise a child. That's
what we had. We were held in what I've come to think of as *the*

basket of the community. It helps to make good the shortcomings of individual families. It provides a container, not tight but semi-permeable, from which to grow and mellow out. Would Donald John have had that, having been raised not in some grassroots urban neighbourhood, but in a household staffed by servants with his dad aggressively out making it in the city? Would he have had folks around him from a tender age to tutor him in ways of diligence and decency? Would he have had much access to a natural environment that could channel any hyperactive tendency into useful occupations, such as carrying in the bags of peat, scything the hayfield or taking out the boat on seas that quickly school the headstrong? 'Fear the sea,' as we were always taught; 'fear' in this sense not as trembling, but as respect.

My sense of the president – America's choice of president – is that he picked up from his mother's side elements of the island's common touch. He picked up the ability to connect with people very deeply at a certain level of their yearning, especially the need unconditionally to belong. But with a disconnect. With smoke and mirrors. With something that's not real, because an adequate foundation was never laid. You see it in those handshakes, a long warm island sort of handshake; but then he plays the power card, yanking people in towards him, pulling them onto his own terms. That's not island. That's not *sana* either, that invites people in. That's the brazen wheeler-dealer of New York. The island will speak of doing things 'for the community'. Donald John does things for himself. He really needs them for himself. And that's the difference from the island's, or, for that matter, from the Melanesian way.

Politically, this isn't about him. This is about what it is in him that connects through to his core constituency. J. D. Vance plays it down when he says that the Scots-Irish 'were day laborers in the Southern slave economy'. The historical research is only now emerging – only now are we developing the confidence to explore our own national shadow – and what we find is that Scots settlers became complicit to the hilt in slavery. The Mackenzies of Kintail, Seaforth and, laterally, Barbados, were big in 'the Africa trade'. They recruited people from back home, not to fill day labourer vacancies, but more as middle management. Racism was the legacy.

You can't oppress and respect people at the same time. If you're not already in a state of inner disconnect, becoming complicit in the machinery of traumatising others will both complete the task and wrap it up in narratives of supremacy, exceptionalism or God-given 'manifest destiny'. As Conrad's Marlow said of Mr Kurtz in *Heart of Darkness*, 'he was hollow at the core'. Soul itself was cleared, or more accurately, put onto the back burner. To the rally call, *Make America Great Again*, we might ask: What makes it not feel great? To the British rally call, *Take Back Control*: At how deep a level, and precisely where, was control really lost?

There is an added element. Donald John says he got his Christian faith from his mother. The hard-line form of Calvinism in which she was raised was brought to Lewis in the 1820s by the heir and daughter of Mackenzie of Barbados. It was propagated in tandem with the Clearances, and subsumed the pre-existing creation, or providentially based spirituality, of a Celtic peoples.[18] The newly imposed evangelicalism emphasised sin, punishment and 'double predestination'. The latter legitimises a fundamental inequality of human beings. God is believed to have set up a binary divide, by which some souls were created to be the Elect or 'saved', and others to be the Damned or 'reprobate', destined to eternal punishment in the fires of hell.

It was the same theology that justified Apartheid in South Africa, black slavery and conquest of the native peoples in America, and which took hold in Papua through missionaries of the Dutch Reformed Church. Predestination is a belief system that, in the harsher forms of its expression, no matter what one does in life, one cannot alter how God cast the die. All of us suffer from the 'total depravity' of sin, some are 'saved', but there is no free will except, and even this is argued, in the ability to accept divine grace. Such a God-ordained presumption of inequality panders perfectly to the psychology of in-groups and out-groups. There are, of course, less religious ways to explain the penchant of Trump and his support base for 'othering'. However, knowing the kind of sermons on which his mother would have been raised, sermons that many a pastor in Trump-supporting parts of America would still preach today, I cannot help but wonder if his obsession with the wall with Mexico is a binary divide that runs right down the middle of his and his supporters' minds. Exit polls on his election showed that 81 per cent of his vote came from white evangelical voters. Most of these have remained loyal, notwithstanding all the scandals. Perhaps it helps a politician to have the doctrine of 'total depravity' as their default. Perhaps it confers a perverse humility by which one can say to another, 'What else can we expect?' If so, no wonder Hilary Clinton's jibe that Trump supporters were a 'basket of deplorables' had such a self-embracing sticking power.

To repeat J. D. Vance again: 'There is a lack of agency here' in this spiritual and material poverty that 'suggests that something else is going on.' We may, he says, have been lucky enough to live the American Dream, that of 'success' and unceasing prosperity, but – and notice the collective ownership with which he says this – 'I want people to understand the American Dream as my family and I encountered it . . . And I want people to understand something I learned only recently: that for those of us lucky enough to live the American Dream, the demons of the life we left behind continue to chase us.'

When Trump was inaugurated, a diminutive piece with an unflattering photograph in the *Stornoway Gazette* announced: 'A man sprung from the loins of a woman from Lewis has taken official control of the most powerful political office in the world.'[19] There was no pride or celebration. The island's evangelical church leadership made plain that his values were unhinged from those of his maternal home. My own view is that Trump's psychological constellation fits like a key into that of his deracinated diaspora constituency, 'Scots-Irish' in archetypal terms. In its wider political psychology, this is no longer a story about Mary Anne, or Donald John, or 'the lonely Isle of Lewis'. This is about a wound that weeps and festers on unhealed across the world. This is how violence, whether suffered or inflicted in its spiral interplay, hollows to the core.

I am left thinking of the work of Yolanda Gampel, an Israeli psychotherapist who spent her life working with survivors of the Holocaust, their children and grandchildren. She describes what she calls the 'radioactive identification' of the trauma set in place by layers of violence. Trauma in this '*Metagama*' sense is psychological injury. It knocks on intergenerationally, blocking the flows of life, she says, like 'indigestible stones, that rattle around the belly of the world'.[20] We might glimpse, then, that colonisation is not just the clearance of the land and, therefore, a blockage in the ability to have a sustained and sustaining relationship with the earth. It is clearance also of the mind. Colonisation of the soul. It leaves behind a harrowed facade, and with it, a cult-susceptible vulnerability to false agency and sense of meaning.[21] In hard business parlance, it levels the ground to a homogenised market surface where the only thing that matters is to have, and keep on having.

Consumption – Can't Get No Satisfaction

This analysis of 'clearance', its extended sense of disconnection, is not intended as a one-size-fits-all explanation of what ails the world. It is offered as a part of a jigsaw, but one that might help to understand our susceptibility to consumerism. Here is what brings my apparent detour full circle to the drivers of climate change. I define consumerism as *consumption in excess of what is needed for dignified sufficiency of living*. While my case study has focused on President Trump as an icon into a bigger picture, that picture is not new.

Other pointers in the direction to which I am nodding include Wendell Berry, the philosopher farmer who wrote *The Unsettling of America*, and whose life's work has been described as based upon the view 'that the modern American way of life is a skein of violence'.[22] In *The Consequences of Modernity*, Anthony Giddens, the leading English sociologist who was an advisor to Tony Blair, described what he called 'the disembedding of social systems'. Here, social relations are lifted out from their local contexts of interaction and restructured 'across indefinite spans of time-space'. To put that in more illustrative language, as the boats rot on the beach, we gradually lose touch with both the characteristics of place and our historical connections with one another.

When that happens, Giddens goes directly on to say, money becomes the substitute for real-life relationships. It becomes 'a medium of exchange which negates the content of goods or services by substituting for them an impersonal standard'.[23] Put more illustratively, it hollows out the soul. He notes that Marx described money as 'the universal whore', language that raises problems today but which can point us to the need for a deeper and more embodied understanding, such as that which is taken by the black feminist writer, Audre Lorde.

Lorde contrasts the pornographic, as mere sensation that is empty of real presence of connection, with what she calls 'the erotic'. To her, the erotic, in a sense of Eros that goes beyond the sexual, represents engagement of the heart. Writing, as she puts it, 'in the face of a racist, patriarchal, and anti-erotic society', she explains: 'Pornography emphasizes sensation without feeling [whereas] ... when I speak of the erotic, then, I speak of it as an assertion of the lifeforce of women; of that creative energy empowered, the knowledge and use of which we are now reclaiming in our language, our history, our dancing, our loving, our work, our lives.'[24]

Note the fullness of life in that description, and so when I speak of consumerism as consumption in excess of dignified sufficiency, I am not advocating a meagre monk-like asceticism. I am not suggesting that we don't have the good things. Quite the contrary. We need to learn the grace of counting blessings in all their cornucopia, otherwise we sabotage that which, as touched on in the previous chapter, I've come to think of as *the cycles of grace*. The ways in which gratitude brings further grace, in contrast to the mean and pinching spirit of crass gracelessness that ends, eventually and somewhere and for someone, in the ennui of bloodletting.

The cut-off point when healthy *consumption* tips over into *consumerism* is when we start to grasp at things with such addictive avidity that we no longer care about the social and environmental relationships embedded in them. That is when economic exchanges turn from blessing to pornography. Perhaps for many social and environmental activists, although they may not have articulated it quite so, this is what makes advanced forms of 'capitalism' so profane. If unrestrained by an ethos of and a requirement for service and accountability to society, such capitalism is no more than the substitution of money for the flows of grace as providence. This

usurpation of the holy accounts for its vulgarity. Therein lies its
vice-like grip – vice in the moral sense – on many if not most of us.

Consumerism, then, comes masquerading as a friend, a balm to
life's afflictions, but is no better than a huckster's snake-oil remedy;
one that enters through the broken skin, spreads toxin and infection, and pollutes the cycles of grace. Consider so-called 'fatigued'
clothing – sandpapered down and ripped by some poor mother in
some goddamned Far East sweatshop, poverty itself appropriated
and glamorised into a garment meant to 'satisfy' fashionable affec-
tations. No wonder, as a great English theologian put it: 'I can't get
no satisfaction / 'cause I try / and I try . . .'

As a corollary of advanced capitalism, consumerism didn't
just arrive from nowhere.[25] Its roots lie mainly in America in the
1920s and 1950s with fears that the economic bubbles built up by
two world wars might collapse with ruinous social effect. In 1925,
Herbert Hoover, later President Hoover, said to a convention of
marketing men: 'The older economists taught the essential influences of "wish", "want" and "desire" as motive forces in economic
progress. You have taken over the job of creating desire.'[26]

A foundation stone was the work of Edward Bernays, a nephew
of Sigmund Freud, who wrote a book in 1928 called *Propaganda* on
the application of wartime arts of persuasion to civilian political
life.[27] Bernays also made his name by persuading would-be liber-
ated women in first-wave feminism to rise up against restrictive
social taboos, handing to American Tobacco the 'goldmine' idea
that cigarettes were 'torches of freedom'. Advertising executives
became known as 'the depth boys', reading the depth psychology
of Freud, Jung and Adler not for healing troubled souls, but to
shift products. The axis of the economy began to shift from a focus
mainly based on satisfying needs, to one that generated wants by
tapping into vulnerabilities in the psyche. Aided by behavioural
psychology, the idea was to 'trigger' buying behaviour by suitably
baiting emotional hooks – freedom, status, power, pride, love, sex,
anger, envy, greed, fear, insecurity and even death – and crafting
adverts that dropped these into the psyche then sat back and waited
for the little fish to bite.

The Hidden Persuaders is a classic study that puts the industry on the couch at a time when the laying of its foundations could be witnessed at first hand, in the decade or so after the Second World War. Vance Packard gives as one of his examples a 1954 advertising conference presentation in Michigan. The agency's remit was from a chemicals company that was wanting to expand the market for hair-perm products. They'd identified a market gap with little girls. With the help of projective tests from child psychology, the agency explored girls' responses to pictures of a child standing at a window. They found – and, perhaps, consistent with the very fashions that they'd helped to forge – that while straight hair was associated with loneliness, curly hair suggested popularity. They also found that little girls' burning concerns were: 'Will I be beautiful or ugly, loved or unloved?' In that case, why didn't they already perm their hair? The mothers were the obstacle. They feared that if their daughters permed their hair, the chemicals might cause damage and that an obsession with looks might make them sexually precocious. How to get around it? The contrast between lonely-straight and happy-curly was packaged into children's TV ads. Pester power became the way to go.[28]

The 'father' of such 'motivational research', known by its critics as 'motivational manipulation', was Ernest Dichter, an admirer and contemporary of Sigmund Freud, born in Vienna in 1907, died in New York in 1991 and named Man of the Year by the Market Research Council as late as 1983. Dichter sold soap for cleansing not just the body but the psyche too. He is often credited with Esso's slogan, 'Put a tiger in your tank', having clicked that petrol could be sold not just as motor fuel, but as personal power. He told the confectionary industry that the British consumed vast amounts of sugar because of their suppressed emotions – a recognition of comfort eating.[29] To the tobacco industry he gave such lines as 'Smoking is fun'. 'Smoking is a reward'. 'I blow my troubles away'. And, to add to that for emotional anaesthesia in a disembedded, disembedding world: 'With a cigarette I am not alone'.[30]

It helps to take an archaeological view, because in the older writers we can watch as it happened. Our school library in Stornoway around 1970 had two hardback copies of John Kenneth Galbraith's book *The Affluent Society.* (I mention that, as an encouragement to all librarians.) Somehow, this study from 1958 intrigued me long before I was able to understand it. But when I did, I was struck by how the great humanitarian economist described what was going on by using the language of addiction:

> The general conclusion of these pages is of such importance for this essay that it had perhaps best be put with some formality. As a society becomes increasingly affluent, wants are increasingly created by the process by which they are satisfied . . . Increases in consumption, the counterpart of increases in production, act by suggestion or emulation to create wants . . . It will be convenient to call it the Dependence Effect.[31]

Most modern marketing books don't even teach this stuff any more. Ask the marketing executives, as I have done in offices of both Saatchi & Saatchi and Collett Dickenson Pearce. If they even know what you're talking about, they'll say: 'We've moved on from all that.' But have they? Might it not have just become part of the woodwork? To be so good at what they're doing, might it be embedded in the disembedded culture of their predecessors' making, unable any longer to see the framing of what they make? As Madonna puts it, we're living in a material world, where the boy with the 'cold hard cash' calls the shots, and she is 'a material girl'.[32]

Partly – these things are always 'partly' because folks are never completely lost – we have eaten of the metaphorical Tree of Knowledge, and are all hooked on its fruit. Our world of nearly 8 billion people is locked in to a fossil fuel economy that drives and is driven by consumption in excess. Given where we're at, technology will have to play a part in the solution. But there was also in the garden, so that we shall 'overcome', a second tree – the Tree of Life – the leaves of which are 'for the healing of the nations', and that in 'the times of restitution of all things'.[33]

How rather splendid, therefore, that Extinction Rebellion's handbook, *This Is Not a Drill*, should have an unexpected and prophetic afterword. Rowan Williams, the former Archbishop of Canterbury, opens his contribution by sounding a note of caution towards the movement's aspirations, a critical friend's whisper of caution, one that I share. 'It might just work,' he says about their plan for revolutionary social transformation, but, 'I can hear the sound of people not holding their breath.' However, as the elder watching over, he doesn't merely leave it there. He pushes on to examine the climate crisis as a deeper sickness in the human condition, one from out of which: 'We may or may not escape a breakdown. But we can escape the toxicity of the mindset that has brought us here. And in so doing we can recover a humanity that is capable of real resilience . . . That's why the creation of panic, with its inevitable accompaniment of self-protection and withdrawal, is useless in addressing the challenge.'[34]

In the place of panic, or perhaps more by bouncing off it, he suggests that we can sow 'the seeds of a future that will offer life – not success, but life', and that this means 'settling to inhabit where we are and who we are'.

An activist of ancient times, one of whom Williams would have roundly approved, told how we have forsaken, 'the fountain of living water, and dug out cisterns . . . cracked cisterns that can hold no water'.[35] While that epitomises the challenge to consumerism, it also invites us to think about the avenues of future advocacy. Perhaps a part of Williams' gentle reservation, his sound of people not holding their breath, is that we may have reached a time when activism itself must ask about its traction and the directions of its calling. Micah White was a young African American activist when he co-created Occupy Wall Street. But as he later provocatively wrote: 'Protest is broken and the people know it worldwide . . . Activism is at a crossroads. We can stick to the old paradigm, keep protesting in the same ways and hope for the best. Or we can acknowledge the crisis [and] embark on . . . a spiritual insurrection . . . a shift away from materialist theories of social change towards a spiritual understanding of revolution.'[36]

That excites me, especially given where it is coming from, because it feels like an unblocking of the greater depth that *satyagraha* calls for in these times. We saw that Gandhi spoke of the need for a constructive programme beyond the obstructive. The obstructive is important. It gets things moving. But on its own has a short shelf life unless there is sufficient latent groundswell to render its effects overwhelming. Although it gets less media coverage, Extinction Rebellion also speaks of taking steps towards a 'regenerative culture'.[37] At the end of a major motorway protest, when he co-founded the GalGael Trust to work on people-centred regeneration in Glasgow, the late Colin Macleod said: 'We've shown them what we're against: now let's show them what we're for.'

What can that look like on the ground, and in a world that urgently must mitigate and adapt to climate change? We have looked here at *clearance*, at inner *collapse* and at the consequences of excess *consumption* at the cutting edge of global carbon emissions. We have seen, with the help of some psychohistory from President Trump's maternal background, how the outer and inner emptiness created is arguably vulnerable to the false satisfiers of consumerism – such as comprise the vanguard of global climate change. Let us now press on towards the seeds of life, towards *community*, and that as Rowan Williams' 'settling to inhabit where we are and who we are'. There will be many openings of the way. I can only share some pointers and from limited experience, but pinned above my desk there are some lines sent by a friend and written by an Irish songwriter:

> After the city we took to the roads and headed West . . .
> We heard an unknown secret music, soft as a fragrance,
> like a glimpse of some fleet-footed animal wild, pure and
> beautiful. We followed.[38]

Community – Pulling Back Life

It's time to go . . . and so it was that the ancient, timeless university where courses are offered in living closed its session late that afternoon in the old village at the head of Loch Leurbost. As the tide began to turn, we waded back across the river, and back up to the road through Murdina Mackenzie's croft. She came out and regaled our passing troupe with Gaelic song. The Papuans responded chorally from out of their own tradition, and so we went upon our way throughout the week, moving from one island neighbourhood to another.

In Scotland we had begun to wake up to the social history that I have shared in the 1970s and 1980s, with the folk rock music of groups like Runrig and world music feeding in. It also helped to have our history for the first time properly told in the run-up to the 1986 centenary of the parliamentary act that gave security of tenure to crofters. Modern land reform started as a faltering trickle in the 1990s. When landed estates came on the market, communities like Assynt, Eigg and Gigha discouraged private buyers – the natives being restless – and pulled together legal structures, funding and the people power to bring their lands back into community ownership.[39]

Often this work was set in wider contexts of what was happening to the world. A number of the leading lights had worked in the global south and had learned from grassroots liberation movements taking place elsewhere. When we launched the Isle of Eigg's land trust in 1991, we put the question: 'Are we really to be fobbed off with the suggestion that lifestyles based on industrial intoxication, nuclear umbrellas, agricultural soil degradation, land expropriation from the powerless and unjust trade relations with the Third World are somehow "viable"?'[40] Today, the Scottish Government has set in place a raft of land reform legislation and a £10 million-a-year fund – resourced by reintroducing business taxes on sporting estates – to help enable community buyouts. Local communities can claim the right of first refusal when land goes on the market. The sale will then proceed, not at speculative market valuations but at a government economic valuation.[41] With over 400 such land trusts now in place, more than half a million acres – nearly 3 per cent of the nation's land – has come under the democratically accountable control of its own residents.

Here we are seeing hybrid ways of management emerging that balance the need for locally elected boards to get on with the day job, but communities engaged in making the big decisions that parallel the style of New England town meetings. There are problems. There are conflicts. There are people! But the scale is such that any resident, usually after a qualifying period, can get involved and build a common future. The main drivers that make these trusts work are having the control of land on which to establish affordable housing, being able then to develop renewable energy resources and enjoying the freedom to get on with entrepreneurial initiatives without a private landowner's approval; and then there is the sheer empowerment of 'can do' – practical, psychological and spiritual – that having at last a supportive framework of Scottish government laws, policies and funding can engender.

With the Papuans that week in April 2019, we visited a variety of such trusts. On the Isle of Harris they've created two new villages, and didn't need to buy the land because they owned it. Neil and Rhoda Campbell at the West Harris Trust told us how their community had been dying. Before the buyout in 2010, the population had declined to 119, and 35 per cent of its houses were either second homes or holiday rentals. Now, with six new affordable and ecologically designed homes built, the resident population has risen to 151, and the number of children in preschool has multiplied from one to seven. They provide the greater part of their own energy with a third of a megawatt capacity from wind, solar and hydro power.[42] The tourists remain welcome, but they now come to a revitalising community and a state-of-the-art visitor centre. Meanwhile, the existence of the neighbouring North Harris Trust helped to draw in private investment that set up a gin and whisky distillery.

Such sustainable development now employs more people than the Lafarge superquarry would have done – over three dozen full-time, and more seasonally. When the COVID-19 virus hit and there was no hand sanitiser on the island, the distillery diverted its production from whisky and gin. Together with a couple of other island businesses, they manufactured the product to WHO specifications and, through a volunteer network, gave it out to the elderly and vulnerable.[43] There we see community resilience in action, restored from the bottom up. Not for nothing does the Isle of Harris Distillery describe itself as 'the world's first social distillery', though that may be a slight exaggeration. Before 1846, when its population of more than 300 was evicted in the Clearances, some resettling as far away as Cape Breton in Nova Scotia, the neighbouring Isle of Pabbay had quite a reputation too – with four stills running when the excise man wasn't looking.[44]

We went to the Pairc Trust that had been established in 2003 and, after a long battle, finally acquired its 25,000 acres from a reluctant private landlord in 2015. Its community of eleven villages and 400 residents had just completed building their first two affordable homes. Like with many of these trusts, women feature strongly in the governance. 'It may be only two houses that we've built so far,' said Fiona Stokes, the manager whom the elected board employs, 'but we'd rather take it at our own pace. Two homes where people stay for twenty years is better than twenty homes where people stay for only two years.' This is what sustainable development can look like on the ground. Ishbel MacLennan, a board member, added: 'People want big solutions to the community's problems, but there are no big solutions. Big solutions don't work. It's the little things that make the difference.'

We introduced the Papuans to Community Land Scotland, the umbrella group that advises how to undertake such buyouts and networks shared experience. Here, David Cameron, a businessman who owns the local garage, said that to turn around a community, four elements are needed:

Political will at local and national government levels;

Technical support to cover any initial gaps in a community;

Financial support to get things going in the early days;

COMMUNITY DESIRE.

His handout had the latter in block capitals. 'We can do it – nothing is off limits,' he concluded. And by the time he'd fired up all the Papuans, because he was so fired up by them, I just wrote across the page: *Wow!*

Change pivots on such desire – on what most sparks life between us – on what Audre Lorde in the sense of Eros called the 'erotic'.

> The dichotomy between the spiritual and the political
> is also false ... for the bridge which connects them
> is formed by the erotic – the sensual – those physical,

emotional, and psychic expressions of what is deepest and
strongest and richest within each of us, being shared: the
passions of love, in its deepest meanings.[45]

The same must go for how we tackle climate change. It is why
we found with land reform that the artists, poets and musicians –
the faith groups and all who live and work with soul – have such
important parts to play in opening up the avenues of inner life
that flow through into outer life. The survival of being, at least
as expressed in the fullness of human being, rests on lubrication
between these two. To say so is not a fluffy platitude. It is a survival
and revival skill. The Spanish Flu of 1918–19 killed 0.6 per cent of
the Norwegian population. Unlike with COVID-19, many were in
their flower of youth. Its unfolding was so well documented by the
country's epidemiologists that a recent study, looking at how societies recover from disasters, used the data as its primary focus. From
this the researchers found that the most robust communities are
'those that build resilience into the social system'. As Hayagreeva
Rao, one of the co-authors, summed up in an interview, when
communities build a diverse base of social cohesion *and* frame their
problems in ways that find common cause rather than blaming
one another, they emerge with strength to solve their long-term
problems. 'First, you need glue to bind a community together,' he
said, 'but you also need WD-40 to reduce friction.'[46]

In this respect, it is striking that the late Elinor Ostrom, the
first woman to win the Nobel Prize in economics, was honoured
for similar lines of thought by the IPCC. *Mitigation of Climate
Change*, the 1,500-page input from Working Group III to its Fifth
Assessment Report (AR5), was dedicated to her in 2014. It cited
what it called her 'fundamental contribution to the understanding
of collective action, trust, and cooperation in the management of
common pool resources, including the atmosphere'.[47] There we
see the science being held in social context. Communities and the
atmosphere are intimately intertwined.

Our Papua delegation moved on down to South Harris, where
we met with Shonny MacAulay, the island's boat builder. We went
and sat beneath the summit that had so very nearly been destroyed.

'A lot came out of that mountain,' he and I will often say to one another. Later in the week we went back up to Lewis and met with island tourism officials. They were fascinated to hear how the Papuans had developed their own ecotourism booking and quality control system, and in so doing, had managed to keep the big hotel developers at bay. And then, before they left the island to go back to Papua, we visited the Point and Sandwick Trust. This generates the best part of a million pounds a year from three wind turbines on their common grazings land. Not only do they pay off their collective carbon karma, but the revenues get poured into all manner of island causes that strengthen the resilience of the surrounding social fabric.

I've simplified this chapter down to four Cs – clearance, collapse, consumption and community. The first three disembed, but the latter re-embeds meaning into life. Often the question arises: What can we do when we wake up and realise that we've got to do something? One answer is to feed the hungry. Refute the graceless spirit of mere utility. Restore the broken cycles of grace, put right the devastation of the past, and celebrate resurgence of the flows of gratitude, of blessedness, of what gives life. On which matter, before we left the Point and Sandwick Trust, the Papuans had an encounter that moved them to the core.

Rhoda Mackenzie and Katie Laing took us to the headland where the *Iolaire* had gone down, the ship that lost 200 men coming home from the horrors of the Great War on New Year's Day in 1919. We looked out across the marker on the rocks, known as *The Beasts of Holm*. The Papuans were very taken by a suggestion that some of the naval ratings would have known that they were on the wrong approach path to the harbour, but such was the authority structure of the times that they would probably not have seen it as their place to challenge the judgement of officers 'from away'.

We stood before the monument, a rope coiled up and cast in bronze. In those days before heated pools, very few islanders could swim. But one man aboard the ship who could so do, took to the water with a line attached. At the first attempt, he got swept under the hull by the force of the backwash. But he resurfaced, and with his knowledge of the sea, waited as he judged aright the rhythm of the waves, and then surged in on a breaking crest to safety on the shore. From there he used the line to pull ashore a hawser. Some forty men got off before the wreck gave a mighty lurch, and the rest were lost.

Something about the sculptor's representation of the rope moved Mama Enggelina very deeply. The heavy hawser coiled round and round at the base. The lighter messenger line that the man had swum with, coiled up above. The chunky Franciscan monk's knot to give it throwing weight, resting steadfastly on top.

Mama was an islander too, some kind of a pastor in the church back in her community. Like Mama Morin, she spoke little in the group, but when she did, everybody quickened to alertness because they knew it would be meaningful. Outwardly she looked incongruous, standing at this tragic location, all zipped-up in a too-large anorak to harbour from the bitter North Atlantic winds. Inwardly, and maybe more than any of us, she was right there. A depth of presence edging to the reverential.

She leaned down, and slowly touched her black-gloved fingers on the coils of bronze. It was as if some instinct sought to bless, or ask as blessing, from a depth of being that survived the *Iolaire*.

'That rope,' she said, and waited for Maria to translate.

'That rope pulled life back into the community.'

THE RAINMAKERS

Early in 2020, it was announced that the decade just ended had been the hottest ever. The Goddard Institute's Gavin Schmidt, who oversees NASA temperature data, said that the world had crossed over into warming of more than 2°F, which is fractionally over 1°C. What's happening is persistent, he said, 'not a fluke due to some weather phenomenon'. There was no doubt as to the reason why. The driver is greenhouse gas emissions.[1]

Let's take stock of where we've got to. In this book I have presented the latest peer-reviewed consensus science. I have shown why such bodies as the IPCC hold carefully to the evidence base in their statements, and why such expert panel reviews are so important in fields of new research that often have little historical baseline data and may produce findings that can lead to disagreement between experts and to contradictory headlines in the media. I am well aware of the limitations of the cautious approach of mainstream science. But I am equally aware of the dangers of not following it. Back in the 1970s, before there was any such body as the IPCC, some scientists suggested, and with some justification, that we were heading towards a new ice age. I can remember as a student in Aberdeen buying an issue of *New Scientist* that reported on the matter. The newspapers of that era ran a rash of articles with headlines like '2 scientists think "Little" Ice Age near', and the *Radio Times* in November 1974 ran with a cover story and apocalyptic imagery: 'The ice age commeth'.[2] It was a scare that came and went. But what if, to assuage public fears, we had started flaring off natural gas as a geoengineering exercise to boost CO_2 emissions and warm the planet? How would that have panned out? In expert matters, there is no substitute for a balance of expertise.

To those who challenge such reliance on consensus science, any qualified person can seek nomination to be a part of the IPCC's writing and reviewing process.[3] It is an open system. For example, we saw that the climate sceptic mathematician, Nicholas Lewis, whose work led to a paper in *Nature* being retracted on account of statistical error, had held precisely such an IPCC position. The range of expert opinions on any given point of emergent science will have outlier positions at both ends. However, if we who are not experts in their fields cherry-pick only scientists who hold optimistic views, or 'gloom pick' only those with pessimistic views, then we will end up with a skewed sense of reality. That's a problem, because if we have political traction, it will follow through into action or inaction that will have ecological and political misleading consequences. Expert panels help us to strike a balance between Cassandra and the boy who cries wolf. If we presume to bypass the need for such balance, we risk the hubris of getting ahead of ourselves. The ego trips up on its shadow, and then those who follow, because they trusted us, will stumble on our vanity too.

For this reason I have rejected both denialist and alarmist interpretations. I recognise that the first is probably mainly motivated by selfishness to sustain opulence, while the second is probably mainly an altruistic concern for others and for other species. There is an asymmetry here. Politically, denialism ramped up by funding from the $5 trillion fossil fuel industry has had an impact far more damaging than alarmism, which mainly limits its harm to unsettling individuals' emotions and sowing confusion about the credibility of real science. Either way, anybody who thinks that the IPCC wilfully overstates or understates the reality can make their own percentage adjustments if they think they have the grounds to do so. It has not been my task to double guess my diverse readers. My task has been, with due caveats, to present the consensus science in summary at face value, and more essentially, given its alarming nature, to ask: *What now?*

What Does it Say to Youth?

The three landmark special reports that we have examined, and SR1.5 especially, are clear that the world must act decisively and fast. I suspect that as the IPCC's Sixth Assessment Report edges towards finalisation in 2022, their language in so saying will harden. But that will depend largely on shifts in the evidence base as the relevant bodies of scientific knowledge expand, and areas of uncertainty, such as the worrying behaviour of clouds under more extreme global warming, become better understood and built into the models.[4]

Irrespective of such uncertainties, climate will remain the most pressing global leadership issue of our time. Wars might kill in the millions and bomb out cities, but those can be rebuilt. Pandemics may decimate but will pass. Climate change may also, directly or indirectly, come to kill in the millions – billions is most unlikely on any credible scenario this century – but flooded cities will never be rebuilt except, perhaps, behind walls, on stilts or floating on pontoons. Urban heritage will be lost forever, like in the ancient deluges described by Plato, where civilisation had to be restored by shepherds from the mountains.[5]

That is why the Paris Agreement places such emphasis on keeping the global temperature rise to below 1.5°C, and even that is problematically high. We have seen how the science can show that this is physically possible. But we have also seen how the physics–politics gap means that nothing adequately adds up. That doesn't mean that the measures being proposed don't have important contributions to make, or that they won't slow down what is coming upon us. It just means that, for now and leaving many things too late, we have the impasse of a 'wicked problem' on our hands.

One response when faced with this is to say that there is no hope: to burrow into a depressive hole, or head off to the hilltops of New Zealand, and to give up trying. Another way, as we've seen denialism move towards, is to carry on consuming fossil fuels and to mortgage the problem onto the next generation. We know what Greta Thunberg rightly thinks of that. A third way, the one that I have most pressed for here, is to pull out all the stops on sustainable development, and that, not in the shrunken sense of 'sustainable growth' into which its critics have tried to appropriate and tame it, but as integral human development. When I speak to senior military officers at staff colleges, I put like this. That if, as they say, 'peace is our business', and if we are concerned with true security, then perhaps we need a shift in mindset as to what constitutes a threat and how resources should be applied to countering it. At the risk of sounding stuck in John Lennon's groove, I say things like: Imagine if, instead of frightening 'enemies' with missiles, we worked to take away the roots of war. Imagine if, in place of dehumanising poverty, we built up the dignity of all. Imagine if, instead of canonising competition, we made a valour of cooperation. Imagine if, when the coronavirus struck, we'd stockpiled silos of hand sanitiser instead. A lot can start to happen when social mores shift, and violence and its symbols are delegitimised.

Curiously, it gets a certain traction because serving soldiers, more than many might expect, have experienced not just what they think of as 'the utility of force', but also the futility of war.[6] All these themes I've outlined are already well set out in international consensus. They're what the Global Goals of the UN's 2030 Agenda for Sustainable Development are all about. As a world, we've planned our work but now must work our plan. And as we've seen, to place priority on doing so will tackle hand-in-hand at least the first of the twin drivers of greenhouse gas emissions – population – and followed through wisely will, I think, erode the roots of feckless consumption too. Some soldiers think that this is off the wall, but most, in my experience, *most* of them, respond thoughtfully.

With land reform and land trusts, I've given some examples of what well-integrated sustainable development can look like on the ground. Here is an approach that works within the market, consistent with a mixed economy. It doesn't demand systemic overthrow in the absence of thought through alternatives, and yet it moderates, and by subtle market spoiling it subverts, monopolistic ownership as a pillar of the capitalist paradigm. Community trusts are not anti-business, as was mentioned with the example of the Harris Distillery – or there's Laig Bay Brewing on Eigg with its

Independence! I.P.A., and there's even non-alcoholic businesses in the islands too! The entrepreneurial spirit can sit very comfortably when it operates answerably within the community. It is not beyond human genius to square the circle between constrained capitalism and communism. We call it communitarianism.

In looking at the lie of the land, we also examined some psychological drivers of insatiable consumerism. A note of caution here. Whenever somebody sums something up in a mnemonic form, such as the four Cs, there'll always be simplification and blind spots. But with that caveat, we saw how the knock-on effects of clearance, collapse and consumption can start to be redressed by rekindling community. Let's share such soft technology around. Let's name the difference that *love* can make when worked into social policy. From the GalGael Trust, my neighbour Gehan Macleod sent out a tweet on Valentine's Day that said: 'Our secret mission is to make LOVE the most common four-letter word in Glasgow.' First amongst the outcomes for the Scottish Government's National Performance Framework is that people should 'grow up *loved*, safe and respected so that they realise their full potential'. This and other objectives are explicitly anchored to the UN's Global Goals which, states Nicola Sturgeon, the First Minister, 'offer a vision of the world that I believe people in Scotland share'.[7]

If we hide such values under a bushel, we won't see or have the legitimacy to cultivate them. We'll miss the chance to work not only with them, but on the 'shadow' sides of our own shortfallings and inevitable hypocrisies. Indeed, the vanity of fearing being rumbled is more a problem than hypocrisy itself. Give me any day an honest hypocrite, than a paragon of virtue seething with resentment and mock humility.

In opening up the realm of metavalues – what Abraham Maslow called 'being values' such as love, beauty and truth, that can only be defined in terms of other higher values – it doesn't mean we merely ice the cake of drab reality. Rather, it helps us not to miss the difference that can make the difference. Often it is the vulnerable, or the guest, who from the underside or an unfamiliar angle will most clearly see the inner nature of a thing. They can put into words what might leave us shy. In 2015, when we had a group of civil servants and elected politicians come from Papua Province to Glasgow, Lewis and Harris, this was their concluding statement to the *Stornoway Gazette* after ten days exploring climate change, the land and community empowerment. It excites me not just what they saw, but what they were able to see:

> We have seen that people here have two things that make their communities work: love, and a sense of ownership. The land in Papua is more productive, but because these people love so much, it holds it all together. They're happy to live for other people and not just for themselves. They understand that land is God's creation.[8]

Real life is messy, and green and social movements are one of those untidy edges. The Buddhist teacher, Joanna Macy, speaks of the need to do 'the work that reconnects'.[9] It is why, when not out protesting on the streets, or sitting in the back rooms drawing up policy proposals, many of those who are engaged in climate change concerns tend to be the backbone of 'think global, act local' initiatives such as community gardens, transition towns, eco-congregations, food banks, co-ops and centres that give hope and also succour to refugees, maltreated and homeless people. So it is that the deep learning called for in our time includes such approaches as collaborative leadership, community building, understanding the UN's sustainable development goals, negotiation, mediation and conflict resolution skills, pathways of nonviolence, and conscientisation – a term that originated from Brazil for the raising of both conscience and consciousness. Oh. And bookkeeping. And technological know-how. And cookery, to so many recipes that reconstitute the world.

Like the person caught outdoors in a blizzard, we must not be paralysed by the scale of that which is. Climate change may be on a biblical scale, but none of us is God. We must be wary of the tendency to over-reach, over-work and burn out. It can help to be humble yet remember Greta Thunberg's maxim: 'No one is too small to make a difference.' Often, what differences we can make in the world will seem to have little direct relevance to climate change.

But if they're faithful to life, everything is interconnected. When our family friend, Paula Cowie, visited and asked what I'd been writing about, she nodded to her teenage son Ollie, with the question: 'But what does it say to him?' What about youth who are frightened for their futures, like in my generation we worried – rightly so, and still with good cause – about the Bomb and nuclear winter? Climate change presents a qualitatively deeper set of concerns, one that, as the thinker Dougald Hine, who co-founded the Dark Mountain project says, renders it 'a category error to try to take up climate change as one topic among others'.[10] Teenagers won't be hanging around for a long list, but here are five suggestions:

Learn the ways of nature and of climate science – Michael Mann runs courses free online.[11]

Learn how to fix things, grow things, cook things, tell stories, and to give and have fun.

Have conversations across the garden fence with real people in real places and in real time.

Build community, and choose work in ways that show the kind of world we want to be.

Become a rainmaker.

In Our Doom Is Our Dharma

Here then, is our precious irony. We stand at a time when the world could slide progressively towards barbarism, towards our common fate. Or we stand at what could be one giant leap forward in human potential. As the Brundtland Report on sustainable development had it, we are at the threshold of *Our Common Future*. What makes this such an important and therefore exciting time to be alive, in spite of everything, is that it is precisely out of the stimulus of our 'doom' that our *dharma* – the opening of the way of life – can emerge.

Never did a dour Scot speak more truly or more joyously, than when Private Frazer of the BBC sitcom *Dad's Army* would declare – 'We're doomed, we're a' doomed' – for in our doom is our *dharma*. If we fully face the just so of reality, reality will open out to us.

Doom means more than merely counsel of despair. Such a sense of final fate, of ruin and of death, came into usage only in the early modern era, from about 1600. The underlying meaning goes back through the Greek to the Proto-Indo-European language family, of which Sanskrit is the closest survivor. Here the root sound, *dhē*, means that which is laid, set or put down. As such, doom means 'law', and to widen from the narrow legal sense, it means the way in which things are ordained or set in place.[12]

From the same root, Russia's law-making assembly is called the Duma. William the Conqueror's survey of land-holding in England and Wales was the Domesday Book. And in Govan, less than a mile downriver from the planned location of COP 26, there used to be – before it was levelled for a shipyard and now a car park – a Doomster Hill, a gathering place from where the old Scots sense of doom as law, as tribal right relationships, would once have been determined.[13] It is in this sense that Shakespeare's Richard III said: 'All unavoyded is the doome of Destiny.'[14] In other words, should we choose not to face reality, reality will choose to face us. *Asatya* will call us back to *satya*. Such is the doom of climate change; a wake-up call to the human condition at this turning point in our biological and cultural evolution.

In a similar vein, the word *dharma* comes from *dher*, also rendered *dhr or dhri*, which is another Proto-Indo-European root and one that shares the same first consonant sound as *dhē*. Although Sanskrit scholars that I have consulted say there is no evidence of a shared origin, the meaning overlaps: this being to hold firmly, to sustain, to support or to nourish. *Dharma* is a central concept in Indian and wider Asiatic spirituality. It is often, if inadequately, translated into English as 'the law' and sometimes as 'truth'.

In the Penguin Classics edition of Hinduism's most sacred text, the *Bhagavad Gita*, Juan Mascaró translates the first two words – *Dharmakshetre kurukshetre* – as 'On the field of *truth*, on the battlefield of life.' From this starting point, and in the dynamic rather than the static legal sense of *dharma*, the cosmic splendour of reality unfolds. Accordingly, the sages say that *dharma* is *satya* in action, that *satya* is the highest *dharma*, and the *Brihadaranyaka Upanishad*, one of the oldest of the Vedic scriptures, says in a passage that is about the creation of the world: 'Verily *dharma* is *satya* (truth): and when a man speaks *satya* they say he speaks *dharma* or when he speaks *dharma* they say he speaks *satya*; thus both are same.'[15]

We might see more clearly now why Mahatma Gandhi insisted that his way of peace and social transformation cannot be instrumental. It cannot be a spiritual method used for worldly ends, no matter how pressing these might be. *Satyagraha* is truth force, soul force or reality force as the way or *dharma* that opens into the divine. Such is its *hieros-arkhein*, its 'hierarchy' in the underlying sense, its holy ordering, its 'enfolded' or 'implicate' order, to borrow from the physicist David Bohm.

When we are 'armed', as Gandhi put it, with this shield, there is no fear in facing doom as *dharma*. 'Yea, though I walk through the valley of the shadow of death, I will fear no evil,' said the Hebrew Psalmist likewise.[16] Facing reality, including the reality of suffering say the Buddhists, is what opens higher consciousness. Such is why our *dharma* lies enfolded in our doom. We listen for 'the unknown secret music', that basic call to consciousness: and maybe, as the playwright Maxim Gorky said, 'All of us are pilgrims on this earth. I've even heard people say that the earth itself is a pilgrim in the heavens.'[17]

Climate change for those of us alive today is where 'into this house we're born / into this world we're thrown'. We don't need to beat each other up about it. But we do need to deal with it. That journey is as much of heart as head, as much of deepening in spirit as of science.

The Rainmaker

If the journey of the head looks like solar panels, heat pumps and green new deals, what of that journey of the heart? How does the *satyagrahi* of our times – the follower of *satya* as reality as the truth of *dharma* – bring influence to bear in the world around them?

In particular, how so the person of no extraordinary power in any outer sense?

Tom Forsyth, the crofter who led me into work with land reform, lived for many of the years that I knew him in a tiny stone hut on a remote peninsula without a road or mains electricity in north-west Scotland.[18] He called his boat *Wu-Wei*, a Chinese term that means action through non-action, and he'd cut a splendid figure in his oilskins at the tiller, sail pulled close into the wind, cruising effortlessly with his cargo for whatever was the land-care task at hand. Between building stone walls, planting trees or tending hives of bees, he'd often pause to discourse on the writings of Carl Jung, and on the Richard Wilhelm translation of an ancient Chinese oracle, the *I Ching*, to which Jung had written a psychological commentary.

Wilhelm had been a missionary scholar in the German colonial port of Tsingtao between 1899 and 1924. He was one of that fine breed of missionaries who seems to have found that what he thought he'd gone to bring had long since preceded him, and so he focused more on learning than on teaching. What makes his *I Ching* translation so important, is that he undertook it with the help of some of 'the most eminent scholars of the old school', who had concentrated in the settlement for safety following the Chinese revolution of 1911.[19] Wilhelm told Jung an eyewitness story, one that the psychologist would tell over and over again, but which, though literally true, should probably not be swallowed too literally.[20]

In the province that surrounded Tsingtao there befell a terrible drought. The grass scorched, the animals were failing, and the people knew that they'd be next. In desperation, they called upon the Protestant missionaries, who came and presumably said their prayers and read their bibles and gave suitably long sermons. No rain.

So then they called the Catholic missionaries, who came and presumably said Hail Marys and prayed with rosary beads and sprinkled holy water. Still no rain.

So they called the traditional Taoist and Confucian priests, who came and lit some joss sticks, and set off guns to frighten away the hungry ghosts that presumably had caused the drought. But not a single drop.

Finally – and interestingly, as the last resort – they called in the Rainmaker. The Rainmaker was a wizened little old man who lived far away. He had to walk for some considerable distance from a neighbouring province. 'What do you need?' they asked when he arrived.

'I need nothing,' he said. 'Just a hut to go and sit.'

After three days, there was an unseasonable fall of snow. It melted and relieved the drought. The peasants soon resumed their normal lives. But Richard Wilhelm, being not just any old scholar but a German Protestant professor, wanted to know exactly what the little old man had done.

'I did nothing,' said the Rainmaker.

'Oh come on,' said Wilhelm. 'Was it magic spells, or incantations, or did you just hit lucky that you only had to wait three days?'

'None of those,' he said honestly.

'Well, what was it then?' demanded the exasperated Wilhelm.

'It's like this,' said the Rainmaker. 'When I was in my home province, my spirit was in the *Tao*, the cosmic harmony. But when I got to this province, I found that it no longer was in the *Tao*.

'So I went and sat inside the hut, and when my spirit settled back into the *Tao*, that's when the clouds began to form.'

The World on You Depends

I am not suggesting that we tackle climate change through non-action. But I am suggesting that there are qualities of presence and inner grounding – whether we call them the *Tao*, *dharma*, *satya* or a triple whammy of 'the way, the truth and the life' – that can awaken the gentleness and visionary motivation that we need. Both action *and* contemplation.

The bush fires in Australia that ushered in the start of 2020 razed more than 10 million hectares, an area that is considerably larger than Scotland. They burnt alive an estimated 1 billion creatures including wildlife, farm animals and pets. With old-growth forests and other long-established habitats gone, some threatened species will probably have met their holocaust of extinction.

When Scott Morrison, the climate change disinterested prime minister, made his embarrassed way back from a holiday in Hawaii, he tried to curry favour by heading down to visit the little town of Cobargo in New South Wales. Fire had swept through its main street, stripping bare the antique wooden facades that once had been an icon of Australia's frontier history. Lost in its path had been the lives of a father and his son who had stayed behind to try and save their home.[21]

The politician's walkabout turned out to be no dreamtime. No rainmaker's hut awaited him in Cobargo. 'You won't be getting any votes down here buddy,' one man heckled.

'How come we only had four [fire] trucks to defend our town . . .?' a woman shouted, as Morrison hurried back to his official car, '. . . because our town doesn't have a lot of money, *but we have hearts of gold.*'

Other fires will come and go, but that woman's testimony had prophetic perspicacity. It cut through the bluster to the living quick. If I lived in a climate change high-risk zone, I'd want to live near her. She came over as a person of no extraordinary power, no special education, and certainly not of 'manners' in the manner to which a prime minister might be accustomed. But she showed the makings of a rainmaker.

What might it mean for us to be rainmakers of today?

In 1957, when he was in his eighties, Carl Jung wrote an extended essay called *The Undiscovered Self*. By the 'self' he meant the soul at levels beyond ego, touching on the ground of deepest being, some might call it that of God within. Writing in the wake of Hiroshima and Nagasaki, and with the Cuban Missile Crisis not far around the corner, the essay opened with the question: 'What will the future bring?' We are living, he suggested, in a time such as the Greeks called *kairos*, a 'metamorphosis of the gods', when a shift takes place in the foundations of humanity.

That tectonic movement will not come from confusing self-knowledge with the knowledge only of our ego-personalities. Neither will it come from sudden mass enlightenment, for such a turning of the spiritual great wheel 'follows the slow tread of the centuries and cannot be hurried'. Rather, given the position at which we stand today, only the individual who is 'anchored in God' can resist 'the physical and moral blandishments of the world'. Put another way, only the individual who is anchored in the great self, in the *satya*, beyond the limited ego consciousness of their small selves, can see through the entrapments of the age. In the vital call to do so, each of us, he said, 'is the makeweight that tips the scales . . . that infinitesimal unit *on whom a world depends*'.[22]

Jung's writings influenced *The Doors*,[23] and one of Jim Morrison's lines in 'Riders on the Storm' seems to echo that just quoted: 'The world on you depends / our life will never end.' That faith means facing up to the shadow, both of ourselves and of our times. It means complementing the outer work of practicality with the inner work, as Jung specified it, requiring both humility and love.

The task is pressing. Another sentence in his essay starkly reveals the stakes: 'Where love stops, power begins, and violence, and terror.' But in their very bleakness, those words define the task of rainmakers today. Of the Mama Enggelinas. Of those who do the science. Of such movements as Greenpeace, WWF, Friends of the Earth, Fridays for Future and Extinction Rebellion's 'ferocious love of these lands'. Of the Australian woman with the heart of gold.

Paulo Freire said: 'This, then, is the great humanistic and historical task of the oppressed: to liberate themselves and their oppressors as well.'[24] There we see the deepest challenge of rainmaking. We have to reach to others as they are, for all we are.

Once Ram Dass, the late Jewish–Hindu spiritual teacher, went on retreat in some lonely ashram miles from nowhere.[25] There he was, for days and days on end, and the only other resident was the sanctuary cat. It would curl up in his lap because 'it loved me, and I loved it'. He'd be meditating there for hours and hours on end, when flip-flop would go the flap, and in would come the cat with a mouse. The only sound to break the silence was the crunching of the little creature's bones.

What to do? What to even think? Such piquancy of nature red in tooth and claw. And out there beyond, out in the world today, the fires and floods and hurricanes – all the galling crises with which climate change and other ills, including COVID-19, beset the world.

Ram Dass could see no grace, no avenue of redemption. Here was all suffering in microcosm. Here was Good Friday's darkness, Christ nailed on the Cross, replayed in endless loop. All that he could do was hold it 'in the light'. And then the framing shifted. Ahhh!

Compassion for the mouse. Compassion for the cat.

The Rainmaker may have been alone in his hut, but everybody knew that he was there. On one side of the scales, the doom of destruction. On the other side, the doom of *dharma*. His make-weight tilt was to themselves. No wonder that they'd called him only as the last resort. In part, the impact of the drought lay in the clouds. In part, it lay within their chosen ways of being. Like in the feeding of the five thousand, once the little boy had shared his bread and fishes, all it took was recognition, blessing, gratitude – and the people's hunger self-resolved.

We miss the deeper truth if we treat the Rainmaker, or Jesus for that matter, as magicians in a literal sense, as mere wonder-workers. The deeper truth points back to who we are, and how that changes our capacity for both mitigation and adaptation in searching circumstances. As with the rope of the *Iolaire* – the Gaelic name means 'eagle' – the Rainmaker pulled life back into the community.

Such is the greatest work we can take on today. A rainmaker's *wu-wei* will not impact directly on climate change. No way! But it may draw out emergence of our deeper humanisation. As I go around, I ask the wisest people that I meet: What can we do, on top of all the obvious outer actions? They mostly say the same, according to their cultures: *metta*, *karuna*, love.

And that strange Hebridean legend that we touched on at the end of our first chapter. The one about the three great floods. What were the storytellers handing down? What signs and portents in their minds? That image of the churches, once standing proud above the sea, submerged, so that 'the pale-faced mermaid, the marled seal and the brown otter shall race and run and leap and gambol – like the children of men at play'. That image, also, of the great Atlantic overflowing: but so the dead will wake up dry, 'Iona will rise on the waters and float there like a crown.'

What might such reference to the symbolic spiritual heart say about the holy ground on which we tread in our daily lives? What might it say about the dead or deadened in us all that could come back to life?

I'll leave it there.

These stories hint how we can be the riders on the storm.

For the mouse, for the cat . . . in every one of us.

Compassion, compassion.

Compassion.

ACKNOWLEDGEMENTS

Hugh Andrew, the managing director of Birlinn, has a knack for nudging authors at the right time. When he asked me in 2019 if I would write an update of my earlier climate change book, *Hell and High Water*, I initially declined. My wife, Vérène Nicolas, had banned me from writing another book after the seven years it took to complete *Poacher's Pilgrimage*. But when she heard of Hugh's suggestion, she encouraged me to undertake it. What could be more necessary in these times?

Initially, I'd planned to take the science as 'given' and not provide an update. Two things changed that. One was that in looking at the spectrum between denial and alarmism, I realised that there was a need to take bearings on what the consensus science currently says. The other was that COP 26 was announced with the intention to take place in the neighbourhood where I live. Vérène has given me the most incredible support in what it took six months to draft. Over that period, I rejected most speaking invitations and worked solidly from dawn to dusk. Socially, I entered COVID-19 lockdown already as a climate hermit. Her steadfast love, and her passion for the task at hand which included critical reading of the drafts, has been a power behind this work.

Hugh, his managing editor Andrew Simmons and my copy-editor James Rose – all with an eye for both literary and technical finesse – have provided not just encouragement, but precious criticism on my drafts. Many generous hands have helped in hidden ways behind the scenes, but I owe a special debt of gratitude to

John and Fiona Sturrock, who went far beyond the calls of friendship, to Maria Latumahina and Adrian Wells, and to Andrew McAulay and John Fellowes. In ways that need not be spelled out, they helped to set and to hold my context. I also thank my sister, Isobel Caplin and her husband Nick, for the family part they played in making my journey lighter.

A range of friends and colleagues have been of great importance as critical readers. I especially thank three climate scientists, Jim Hansom, David Armstrong McKay and Graeme MacGilchrist, for looking at selected specialist aspects of my writing. I emphasise that any remaining failings of understanding are mine, and I welcome notification of any readers' corrections that might be incorporated into any later editions of this work. For general reading or for incisive comments over many discussions, I am hugely grateful to John Ashton, Matt Carmichael, Ian Christie, Paula Cowie, Luke Devlin, Timothy Gorringe, Laura Hope-Gill, Miki Kashtan, Murdo Macdonald, Babs Macgregor, Gehan Macleod, Duncan McLaren, Alison Phipps and Robert Swinfen. They are not responsible for points where I may not have agreed with their suggestions, or for errors that I may unwittingly have introduced.

Institutionally, I am grateful to the GalGael Trust for temporarily relieving me of certain duties, to the Centre for Human Ecology that has held a local grounding, and to the University of Glasgow whose granting of an honorary professorial role provided me with access to the research materials and collegiality of which my readers will share the benefit.

Lastly, my warm thanks to you, my readers. I had hoped to offer something half this length, but to get beyond polemic, and to honour the diversity of my task, I failed. And yet, says Nan Shepherd in *The Living Mountain*: 'To know . . . with the knowledge that is a process of living . . . is not done easily nor in an hour. It is a tale too slow for the impatience of our age, not of immediate enough import for its desperate problems. Yet it has its own rare value.'

Shepherd would have known storm riding as a rare survival skill. And then, with Eliot: 'fare forward, voyagers'.

GLOSSARY OF ACRONYMS

AFOLU	Agriculture, Forestry and Other Land Use
BECCS	Bioenergy with Carbon Capture and Storage
CDR	Carbon Dioxide Removal
CO_2	Carbon dioxide, the principal persistent greenhouse gas
CO_2eq	Includes other greenhouse gases as *equivalents* (also CO_2e or COE)
COP	Conference of the Parties (national governments)
COVID-19	Coronavirus disease 2019
IPCC	The Intergovernmental Panel on Climate Change
NGO	Non-Governmental Organisation (usually a not-for-profit or a charity)
ppm(b)	Parts per million (or billion)
RCP	Representative Concentration Pathway (greenhouse gas concentrations)
SR1.5	IPCC Special Report, *Global Warming of 1.5°C*
SRCCL	IPCC Special Report, *Climate Change and Land*
SROCC	IPCC Special Report, *The Ocean and Cryosphere in a Changing Climate*
SSP	Shared Socioeconomic Pathway
UN	The United Nations
WHO	World Health Organization

NOTES

All URLs were valid as of summer 2020. Full links can be viewed by pasting the shortened ones into a browser. Beware case sensitivity and alphanumeric ambiguity of zeros and ones. To minimise clutter, quotations from IPCC special reports are not endnoted, but can be found by searching within the specified report. A webpage for this book will be maintained at http://www.alastairmcintosh.com/ridersonthestorm.htm.

Chapter 1 – A Walk along the Shore

1   Weston Price, 'Isolated and Modernized Gaelics', *Physical Degeneration: A Comparison of Primitive and Modern Diets and Their Effects*, Chapter 4, 1938: http://bit.ly/2TmjEI6.

2   Andy Moir, 'Development of a Neolithic pine tree-ring chronology for northern Scotland', *Journal of Quaternary Science*, 27:5, pp. 503–8, 2012: http://bit.ly/2VoQw5D. The pines at Leurbost have not been dated, my discussion is illustrative only.

3   Johanna Lehne and Felix Preston, 'Making Concrete Change: Innovation in Low-carbon Cement and Concrete', Chatham House, 13 June 2018: http://bit.ly/2Tiqfn8.

4   Roland Jackson, 'Eunice Foote, John Tyndall and a question of priority', *Notes Rec. Royal Society*, 74, 2019, pp. 105–18: http://bit.ly/32J2EjF.

5   David McKay, 'Fact-Check: do tipping points commit us to rapid catastrophic warming?', climatetippingpoints.info, 15 April 2019: http://bit.ly/2Pzogd2.

6   'Carbon Dioxide: Vital Signs of the Planet', NASA, 2020: https://go.nasa.gov/2weu9FC.

7   Hannah Ritchie and Max Roser, 'CO$_2$ and Greenhouse Gas Emissions',

How did CO_2 emissions change over time?, Our World in Data, update December 2019: http://bit.ly/2VxN48R.

8   Ritchie and Roser, 'CO_2 and Greenhouse Gas Emissions', Cumulative CO_2 emissions: http://bit.ly/2Tonloo.

9   Ritchie and Roser, 'CO_2 and Greenhouse Gas Emissions', Consumption-based (trade-adjusted) CO_2 emissions (figures show by hovering over map): http://bit.ly/2wYrMXX.

10   Ritchie and Roser, 'CO_2 and Greenhouse Gas Emissions', Per capita CO_2 Emissions: http://bit.ly/389SYQg.

11   *UK's Carbon Footprint 1997–2016*, UK DEFRA, 2019, p. 3: http://bit.ly/2sNkPHv.

12   '800,000 Years of Carbon Dioxide', Climate Central, 1 May 2019: http://bit.ly/2T9M2OL.

13   NASA, 'Carbon Dioxide'.

14   'Global Greenhouse Gas Emissions (by economic sector)', EPA, 2014: http://bit.ly/2VA47HI.

15   *UK Energy Mix 2017*, UK DBEIS, 2018, p. 27: http://bit.ly/36edoYP.

16   'UK Energy and Emissions', Energy & Climate Intelligence Unit, figures for 2015: http://bit.ly/2PALgse.

17   'The NOAA Annual Greenhouse Gas Index (AGGI), 2019', *NOAA*, table 2, 2019: http://bit.ly/2x126Ki. NOAA presents the data in a form for public consumption. As the underlying science is complicated, differing figures may be encountered. See Chapter 8 in the IPCC's AR5 Working Group 1 contribution, 2013, especially p. 661, on industrial era radiative forcing: http://bit.ly/3ai5RsX.

18   Ed Hawkins, 'Defining Pre-Industrial', Climate Lab Book, 25 January 2017: http://bit.ly/2TpsnLb.

19   Kevin Edwards, K. D. Bennett and Althea Davies, 'Palaeoecological perspectives on Holocene environmental change in Scotland', *Trans. Royal Society of Edinburgh*, 110:1–2, 2019, pp. 199–217: http://bit.ly/39mko8H. See also, on prehistoric Scottish pinewoods: http://bit.ly/3arvEis.

20   M. James Salinger, 'Agriculture's influence on climate during the Holocene', *Agricultural and Forest Meteorology*, 142:2–4, 2007, pp. 96–102: http://bit.ly/2VPe83T. This paper suggests that the rise began 8,000 years ago. I have used Ruddiman's 7,000-year figure (see below).

21   Matthew Brander, 'Greenhouse Gases, CO_2, CO_2e, and Carbon: What Do All These Terms Mean?', Ecometrica, 2012: http://bit.ly/39r17q80.

22   'Understanding Global Warming Potentials', EPA, undated: http://bit.ly/39xWZQh. Also, on William Ruddiman: Daniel Headrick, 'Global Warming, the Ruddiman Thesis, and the Little Ice Age', *Journal of World History*, 26:1, 2015, pp. 157–60: http://bit.ly/39yIxHm.

23   William Ruddiman, *Plows, Plagues & Petroleum: How Humans Took Control of Climate*, Princeton University Press, Princeton, NJ, 2005. See also: William Ruddiman, 'The early anthropogenic hypothesis: Challenges and responses', *Reviews of Geophysics*, 45:4, 2007: http://bit.ly/2IgVWYN.

24   Kelly April Tyrrell, 'Ancient farmers spared us from glaciers but profoundly changed Earth's climate', ScienceDaily, 6 September 2018: http://bit.ly/3aowaoK. This is a plain-language report that explains the *Nature Scientific Reports* paper on marine isotopes: https://go.nature.com/3900nzN.

25   'Sea Level – Vital Signs', NASA, 2019: https://go.nasa.gov/2uRTrsN. The IPCC report (next reference) gives the rate of increase for 2005–15 as 3.6 mm.

26   Special Report SROCC, IPCC, A3.1, p. SPM-10: http://bit.ly/2Tz47ow.

27   That is, assuming that global rates are applicable regionally, which is not always so. And this is a back-of-envelope figure taken off such graphs as 'Sea Level Rise', CSIRO, 2017: http://bit.ly/3aupBdg.

28   Judith Wolf, 'Coastal flooding: impacts of coupled wave–surge–tide models', *Natural Hazards*, 49, 2008, pp. 241–60: http://bit.ly/2IiVvNz.

29   See account of my mother's terrifying experience that night in the Introduction to my earlier book on climate change, *Hell and High Water*, Birlinn, Edinburgh, 2008.

30   Bruno Castelle, et al., 'Increased Winter Mean Wave Height, Variability, and Periodicity in the Northeast Atlantic Over 1949–2017', *Geophysical Research Letters*, 45:8, 2018, pp. 3586–96: http://bit.ly/32MNqtS. There is, however, evidence from salt in Greenland ice cores that complicates the picture, suggesting that extreme storms in the North-east Atlantic have not increased in frequency when looked at over the past 600 years: see Alastair Dawson, Sue Dawson and William Ritchie, 'Historical Climatology and coastal change associated with the "Great Storm" of January 2005, South Uist and Benbecula, Scottish Outer Hebrides', *Scottish Geographical Journal*, 123:2, 2007, pp. 135–149: https://bit.ly/2VQCCrw.

31   John Coll, 'Sensitivity of Ferry Services to the Western Isles of Scotland to Changes in Wave and Wind Climate', *Journal of Applied Meteorology and Climatology*, 52:5: http://bit.ly/2wtNOkS.

32   'Seafarer hands over the tiller after completing 45-year career', *Stornoway Gazette*, 2 January 2020, p. 8.

33   A. F. Rennie and J. D. Hansom, 'Sea level trend reversal: Land uplift outpaced by sea level rise on Scotland's coast', *Geomorphology*, 125:1, 2011, pp. 193–202.

34   For example, in the work of Professor Stewart Angus of Stornoway, as in Murray Macleod, 'Coastal areas under threat from rising seas and worsening weather', *West Highland Free Press*, 25 October 2019, p. 2.

35   Hasanuddin Zabidin, et al., 'Land subsidence of Jakarta (Indonesia) and its relation with urban development', *Natural Hazards*, 59:3, 2011: http://bit.ly/39nqHHl.

36   'Coastal Risk Screening Tool', Climate Central: http://bit.ly/2uFoB2Y. Thanks to Professor Frank Rennie of the University of the Highlands and Islands for finding me this GIS resource.

37   Genesis 1:2 and 6:5.

38 Otta Swire, *The Outer Hebrides and Their Legends*, Oliver & Boyd, London, pp. 63–5.

39 Alexander Carmichael, *Carmina Gadelica*, vol. 2, T. & A. Constable, Edinburgh, 1900, pp. 270–2: http://bit.ly/2uVKNcU.

40 Swire, *The Outer Hebrides*. For further discussion and sources see my *Poacher's Pilgrimage: An Island Journey*, Birlinn, Edinburgh, 2016, pp. 144–5 and 354–5.

Chapter 2 – Impacts on the World of Ice and Oceans

1 Michael Mann, Twitter, 21 January 2020: http://bit.ly/2x4vM9b.

2 Naomi Oreskes and Erik M. Conway, *Merchants of Doubt: How a Handful of Scientists Obscured the Truth on Issues from Tobacco Smoke to Global Warming*, Bloomsbury, London, 2012.

3 Donald Trump, Twitter, 6 November 2012: http://bit.ly/2PJEPmu.

4 'About the IPCC', IPCC, undated: http://bit.ly/2x9CRFG.

5 The IPCC assessment reports are: AR1 or FAR (First AR), 1990; AR2 or SAR (Second AR), 1995; AR3 or TAR (Third AR), 2001; AR4 (Fourth AR), 2007; AR5, 2014; and pending, AR6 due in full in 2022.

6 'IPCC AR6 WGIII (Mitigation) Report Schedule', 5 November 2018: http://bit.ly/2x21bcm.

7 Zeke Hausfather, et al. (inc. Gavin Schmidt), 'Evaluating the Performance of Past Climate Model Projections', *Geophysical Research Letters*, 47:1, 2019: http://bit.ly/32NWH53.

8 Stefan Rahmstorf, et al., *The Copenhagen Diagnosis*, Elsevier, 2009: http://bit.ly/2uWlCHc.

9 Bidisha Banerjee and George Collins, 'Anatomy of IPCC's Mistake on Himalayan Glaciers and Year 2035', Yale Climate Connections, 4 February 2010: http://bit.ly/2TmDN24.

10 'AR6 Synthesis Report: Climate Change 2022' (showing report structure), IPCC, 2019: http://bit.ly/38puKBE. IPCC reports index page is at: https://www.ipcc.ch/reports/.

11 'What is the United Nations Framework Convention on Climate Change?', United Nations, undated: http://bit.ly/3ctKIhx.

12 *Daily Record*, 'City of Glasgow contains the 10 most deprived areas in Britain', 9 July 2018: http://bit.ly/2TonCRT.

13 UK Government, 'COP 26 Postponement', press release, 1 April 2020: https://bit.ly/2waOoo4.

14 My references are to the entire text downloads on the IPCC's website, rather than the individual chapter links for each report, also given on the website. The full links I've used are: **SROCC** https://report.ipcc.ch/srocc/pdf/SROCC_FinalDraft_FullReport.pdf; **SRCCL** https://www.ipcc.ch/site/assets/uploads/2019/08/Fullreport-1.pdf; **SR1.5** https://www.ipcc.ch/site/assets/uploads/sites/2/2019/06/SR15_Full_Report_Low_Res.pdf.

15 Michael Mastrandrea, et al., 'Guidance Note for Lead Authors of

the IPCC Fifth Assessment Report on Consistent Treatment of Uncertainties', IPCC, 2010: http://bit.ly/3csO8B3.

16 'DoD News Briefing – Secretary Rumsfeld and Gen. Myers', US Department of Defense, 12 February 2002: http://bit.ly/2PL213E.

17 I've quoted from the pdf versions of the IPCC special reports as given in **bold** above.

18 Published September 2019, see **SROCC** permutations at: https://www.ipcc.ch/srocc/home/. All my SROCC quotations unless otherwise stated are from: https://report.ipcc.ch/srocc/pdf/SROCC_FinalDraft_FullReport.pdf.

19 *Global Energy and CO₂ Status Report 2019*, International Energy Authority, 2019: http://bit.ly/3arZTWw.

20 UN Environment Programme, *Emissions Gap Report 2019*, Nairobi, December 2019: http://bit.ly/3arVmDz.

21 'Climate Updates: What have we learnt since the IPCC 5th Assessment Report', The Royal Society, London, 2017: http://bit.ly/2TEG7jJ.

22 'Fracking prompts global spike in atmospheric methane, study suggests', ScienceDaily, 14 August 2019: http://bit.ly/2VJG5K4.

23 Peter Wadhams, 'Arctic Ice Cover, Ice Thickness and Tipping Points', *AMBIO*, 41, 2012, pp. 23–33: http://bit.ly/2x9V5a6.

24 Natalia Shakhova, et al. (inc. Igor Semiletov), 'Ebullition and storm-induced methane release from the East Siberian Arctic Shelf', *Nature Geoscience*, 7, 2014, pp. 64–70: https://go.nature.com/2TAfj4d.

25 Antoine Berchet, et al., 'Atmospheric constraints on the methane emissions from the East Siberian Shelf', *Atmospheric Chemistry and Physics*, 16, 4147–57, 2016: http://bit.ly/2IhTUrh.

26 Examples are Arctic News and the Arctic Methane Emergency Group (AMEG), which campaigns for pre-emptive geoengineering. See also part of my Crisis Forum debate with AMEG members archived at http://bit.ly/32NkaU5.

27 David Archer, 'Much ado about methane', RealClimate, 4 January 2012: http://bit.ly/32M81yH.

28 David A. McKay, 'Fact-Check: is an Arctic "Methane Bomb" about to go off?', climatetippingpoints.info, 13 May 2019: http://bit.ly/39lsFbe.

29 See reply in 2012 to Dr John Nissen of the Arctic Methane Emergency Group from Sir John Bebbington, the UK government's chief scientist: http://bit.ly/2TjB2OW.

30 Richard Poore, Richard Williams and Christopher Tracey, *Sea Level and Climate*, US Geological Survey, Fact Sheet 2, 2011: https://on.doi.gov/2vCwmL6.

31 Donald Trump, Twitter, 18 January 2020: http://bit.ly/2TzHIHS.

32 John Schwartz and Richard Fausset, 'North Carolina, Warned of Rising Seas, Chose to Favor Development', *New York Times*, 12 September 2018: https://nyti.ms/3asCm88.

33 'Percent change in acidity', NOAA PMEL Carbon Program, undated: http://bit.ly/2VI7Q5P.

Chapter 3 – Climate Change on Land and Human Life

1   Published August 2019, see **SRCCL** permutations at: https://www.
     ipcc.ch/report/srccl/. All SRCCL quotations unless otherwise stated
     are from: https://www.ipcc.ch/site/assets/uploads/2019/08/Fullreport-1.
     pdf.
2   IPCC, Twitter, 9 August 2019: http://bit.ly/3avJBw1.
3   David Ussiri and Rattan Lal, 'The Role of Nitrous Oxide on Climate
     Change' in *Soil Emission of Nitrous Oxide and its Mitigation*, Springer,
     2012, pp. 1–28: http://bit.ly/2PJarsE.
4   'UK Climate Averages', Met Office, 2010: http://bit.ly/2TGyZDw.
5   Giulio Corsa, 'There is no evidence that "global warming" was
     rebranded as "climate change"', The Conversation, 12 March 2020:
     https://bit.ly/3ahOBol.
6   Stefan Rahmstorf, 'Q&A about the Gulf Stream System slowdown and
     the Atlantic "cold blob"', RealClimate, 14 October 2016: https://bit.
     ly/3bu5Ym1.
7   G. P. Wayne, 'Global warming vs climate change', Skeptical Science,
     update 8 January 2017: https://bit.ly/2Jg8TTi.
8   Damian Carrington, 'Plummeting insect numbers "threaten collapse of
     nature"', *The Guardian*, 10 February 2019: https://bit.ly/2JeQ173.
9   Callum J. Macgregor, et al., 'Moth biomass increases and decreases over 50
     years in Britain', *Nature Ecology & Evolution*, 3, pp. 1645–9, 2019: https://
     go.nature.com/2JdeZnr. See also summary report: Shelley Hughes,
     'Scientists find no evidence for "insect Armageddon" but there's still cause
     for concern', Phys.Org, 12 November 2019: https://bit.ly/2xqHSto.
10   Caspar Hallmann, et al., 'More than 75 percent decline over 27 years in
     total flying insect biomass in protected areas', PLOS ONE, 18 October
     2017: https://bit.ly/33Z1YYc. I have to say that I could not help but
     wonder why this German study was published in a generalist open-
     access journal rather than one with specialist professional authority.
11   R. Fox, et al., 'The State of Britain's Larger Moths 2013', Butterfly
     Conservation and Rothamsted Research, 2013: https://bit.ly/3bl3Z3k.
12   Human Biosecurity, 'Aircraft Disinsection – Information for Travellers
     on International Aircraft', Department of Health, Australian
     Government, undated: https://bit.ly/2UBOJZ1.
13   'Quantitative risk assessment of the effects of climate change on selected
     causes of death, 2030s and 2050s', World Health Organization, 2014,
     pp. 60, 61: https://bit.ly/2UEvhej.
14   Peter Dockrill, 'There's Something Special About Bat Immunity That
     Makes Them Ideal Viral Incubators', Science Alert, 12 February 2020:
     https://bit.ly/2Upe2P5.
15   Ben Westcott and Shawn Deng, 'China has made eating wild animals
     illegal after the coronavirus outbreak. But ending the trade won't be
     easy', CNN, 6 March 2020: https://cnn.it/3aqknPL.
16   Jem Bendell, 'The Climate for Corona – our warming world is more

vulnerable to pandemic', Professor Jem Bendell blog, 23 March 2020: https://bit.ly/2QPAqzb.

17 World Health Organization, 'WHO checklist for influenza pandemic preparedness planning', 2005: https://bit.ly/2yeGH0P.

18 Brad Zarnett, 'Is Covid-19 the Silver Bullet For a Stable Climate?', Medium, 20 January 2020: https://bit.ly/2xspDnG.

19 Genevieve Guenther, Twitter, 25 February 2020: https://bit.ly/2R5Ttp5.

20 Yilei Sun and Norihiko Shirouzu, 'China's auto industry wastes no time coaxing drivers back to showrooms after lockdown', *Reuters*, 27 March 2020: https://reut.rs/2UC6A1S.

21 Oliver Milman and Emily Holden, 'Trump administration allows companies to break pollution laws during coronavirus pandemic', *The Guardian*, 27 March 2020: https://bit.ly/3apjjvM.

22 Alex Trembath and Seaver Wang, 'Why the COVID-19 Response Is No Model for Climate Action', The Breakthrough Institute, 20 March 2020: https://bit.ly/2WNAgMd.

23 Doyle Rice, 'Planet is "way off track" in dealing with climate change, UN report says', *USA Today*, 12 March 2020: https://bit.ly/2JiJI2l.

24 Nicholas de Jong, 'MacKintosh, John', *Dictionary of Canadian Biography*, 1982: http://bit.ly/3aqyOTD.

25 '2014 Climate Change Adaptation Roadmap,' Department of Defense, Foreword: http://bit.ly/2TlDhkL.

26 Lina Eklund and Darcy Thompson, 'Is Syria really a "climate war"?', The Conversation, 21 July 2017: http://bit.ly/32PlOUZ.

27 Jordan Evans, 'Hurricane tag team of Rosa and Sergio set rainfall records for October', *Cronkite News*, 17 October 2019: http://bit.ly/32Me6Lx.

28 William Lacy Swing (director-general), 'Foreword', *World Migration Report 2018*, UN International Organization for Migration, 2018: http://bit.ly/38hGSok.

29 Heather Stewart and Rowena Mason, 'Nigel Farage's anti-migrant poster reported to police', *The Guardian*, 16 June 2016: http://bit.ly/2IkP1O8.

30 Per Espen Stoknes, *What We Think About When We Try Not To Think About Global Warming*, Chelsea Green, Hertford, VT, 2015, pp. 56–7.

31 'Controversy over the IPCC's 1.5°C Special Report (SR1.5)', Third World Network, 2 July 2019: http://bit.ly/32OMsxd.

32 Zeke Hausfather, 'Explainer: How "Shared Socioeconomic Pathways" explore future climate change', Carbon Brief, 2018: http://bit.ly/2wobCXv.

33 CNN, 'Watch Mnuchin and Lagarde clash over the climate crisis', *CNN News*, 24 January 2020: https://cnn.it/2Ifh7dQ.

34 UN General Assembly, Report of the World Commission on Environment and Development, 1987, p. 24, 3.27: http://bit.ly/2TFntIL.

35 'Develop' as from 'development', Online Etymology Dictionary: http://bit.ly/3cusBrA.

36 UN Declaration on the Rights of Indigenous Peoples, United Nations, 2007: http://bit.ly/38qQPQd.

37   Aldo Leopold, *A Sand County Almanac*, Penguin Classics, London, 2020, p. 262.

Chapter 4 – Containing Global Warming to Within 1.5°C

1   'Paris Agreement: Status of ratification', United Nations, 4 March 2020: http://bit.ly/38hIVZy.

2   Jonathan Easley, 'Trump cements "America First" doctrine with Paris withdrawal', The Hill: http://bit.ly/2IjTReA.

3   'Paris Agreement', United Nations, 2015: http://bit.ly/2wwmiDi.

4   Published October 2018, see **SR1.5** permutations at: https://www.ipcc. ch/sr15/. All my SR1.5 quotations unless otherwise stated are from: https://www.ipcc.ch/site/assets/uploads/sites/2/2019/06/SR15_Full_ Report_Low_Res.pdf.

5   In SR1.5, SPM-C, p. 12. I have omitted statements of interquartile ranges.

6   Lisa Cox and Nick Evershed, '"It's heart-wrenching": 80% of Blue Mountains and 50% of Gondwana rainforests burn in bushfires', *The Guardian*, 16 January 2020: http://bit.ly/2PMASNZ.

7   Christopher Flavelle and Patricia Mazzei, 'Florida Keys Deliver a Hard Message: As Seas Rise, Some Places Can't Be Saved', *The New York Times*, 4 December 2019: https://nyti.ms/2VIgG3v.

8   Matthew Yglesias, 'An expert's case for nuclear power', *Vox*, 28 February 2020: http://bit.ly/383v8Ws.

9   Dan Yurman, 'Next generation nuclear: 25MW, smaller, safer, can be sited anywhere', energypost.eu, 10 April 2019: http://bit.ly/2VIcktl.

10   Mark Nelson, 'German electricity was nearly ten times dirtier than France's in 2016', Environmental Progress News, 12 September 2019: http://bit.ly/2TzC7kE.

11   Adele Peters, 'These Enormous Fans Suck CO2 Out Of The Air And Turn It Into Fuel', *FastCompany*, 21 September 2015: http://bit. ly/2VJHRvi.

12   David Keith, et al., 'A Process for Capturing CO_2 from the Atmosphere', *Joule*, 2:8, pp. 1573–94, 2018: http://bit.ly/2If8IXI.

13   Taking Keith's figures at face value and a median cost of $163, let's round that to £130 per ton of captured, compressed, CO_2. If 1,000 litres of petrol emits about 2.3 tons of CO_2, multiplied by £130 that gives a cost of £299 per thousand litres; or, about 30p to chase after and retrieve the tailpipe emissions of CO_2 from each litre of petrol. Thanks to @ThatcherUlrich for checking this calculation.

14   David Keith, Twitter (edited), 25 October 2019: http://bit.ly/2PMzY3W.

15   'Greenhouse Gas Removal', The Royal Society and The Royal Academy of Engineering, London, 2018: http://bit.ly/2uTsAwD.

16   'Negative Emissions Technologies: What role in meeting Paris Agreement targets?' European Academies' Science Advisory Council,

EASAC Policy Report 35, Halle (Germany), 2018, pp. 9–10 and Annex 7: http://bit.ly/38sR8KI.

17   Giulia Realmonte, et al., 'An inter-model assessment of the role of direct air capture in deep mitigation', *Nature Communications*, 10, 22 July 2019: https://go.nature.com/38lFjpv.

18   Duncan McLaren, Twitter, 31 October 2019: http://bit.ly/2uUBeej.

19   Duncan McLaren and Nils Markusson, 'The co-evolution of techno-logical promises, modelling, policies and climate change targets', *Nature Climate Change*, 2020, at press.

20   Scotland has traditionally been a world leader in such technologies. Biofuel from barley is filtered through the kidneys for its warming effect. This reduces the need for home heating and thereby cuts the domestic carbon footprint. A twelve-year-old Caol Ila does it very nicely.

21   Dave Keating, 'EU Labels Biofuel From Palm Oil As Unsustainable, Bans Subsidies', *Forbes*, 14 March 2019: http://bit.ly/2TjNVsi.

22   Pete Smith and Pep Canadell, 'Removing CO_2 from the atmosphere won't save us: we have to cut emissions now', The Conversation, 7 December 2015: http://bit.ly/2uWEJ3U. I have quoted from an author article; the full *Nature* article is at: https://go.nature.com/32NcmBw.

23   Kevin Anderson and Glen Peters, 'The trouble with negative emissions', *Science*, 354:6309, 2016, pp. 182–3: http://bit.ly/3cumZOb.

Chapter 5 – Sceptics and the Psychology of Denial

1   'Constitution of the Independent State of Papua New Guinea', 1975, p. 2: http://bit.ly/2TqcMel.

2   'The international response to climate change: A history', UNEP, 2000: http://bit.ly/3cvbq9C.

3   'Institute of Public Affairs', Tobacco Tactics (wiki): http://bit.ly/2TDhOmw.

4   'Institute of Public Affairs (IPA)', Desmog, 2016: http://bit.ly/32ThC6W.

5   Clive James, 'Mass Death Dies Hard', in Jennifer Marohasy (ed.), *Climate Change: The Facts 2017*, Institute of Public Affairs, Melbourne, 2017, pp. xv, 320, 334–5.

6   Mark O'Connell, 'Why Silicon Valley billionaires are prepping for the apocalypse in New Zealand', *The Guardian*, 15 February 2018: http://bit.ly/2ww3rsI.

7   Peter Taylor, *Chill: A reassessment of global warming theory*, Clairview, East Sussex, 2009, pp. 232, 268–9, 301. The *ECOS* debate in 2010 has since been lost in a website revamp. I retain the email thread.

8   Emails from Peter Taylor drawn upon here are 31 October 2010 and 18–19 November 2019.

9   'Board of Trustees', Global Warming Policy Foundation, 3 February 2020: http://bit.ly/2x4HUHi.

10 Bob Ward, 'Secret funding of climate sceptics is not restricted to the US', *The Guardian*, 15 February 2013: http://bit.ly/2vCPcSk.

11 Richard Collett-White, 'Major Tory Donor and Johnson Backer to Lead UK Climate Science Denial Group', DeSmog UK, 18 December 2019: http://bit.ly/2VNrKwj.

12 HowTheLightGetsIn panel, Caspar Melville chairing Benny Peiser, Rupert Read and Alastair McIntosh, 1 June 2013, video by Paul Swann: http://bit.ly/2TDtwNL.

13 This is all data logged for the first five years, together with my article in *Reforesting Scotland*: http://bit.ly/2x7BdnT.

14 Hannah Devlin and Robin Pagnamenta, 'Major cities at risk from rising sea level threat', *The Times*, 1 December 2009: http://bit.ly/2x1FCIX.

15 'Andrew W. Montford', Desmog, 2017: http://bit.ly/3cy6T63.

16 'New Paper: UK Media Coverage Of Shale Gas Is "Hopelessly Biased"', Global Warming Policy Foundation, 28 September 2018: http://bit.ly/3aDahuX.

17 Tamino, 'The Montford Delusion', RealClimate, 22 July 2010: http://bit.ly/3aqMTAt.

18 Alastair McIntosh, 'Review of *The Hockey Stick Illusion*', *Scottish Review of Books*, 6:3, August 2010: http://bit.ly/2VIJ45C.

19 Bishop Hill, 'Scottish Review of Books', 14 August 2010: http://bit.ly/3apYCiQ; and 'Did he read it?' 17 August 2010: http://bit.ly/2IgwuT5.

20 For the latest work on this see Raphael Neukom, et al., 'No evidence for globally coherent warm and cold periods over the preindustrial Common Era', *Nature*, 571, 24 July 2019, pp. 550–4: https://go.nature.com/2IoU5Rz.

21 Tamino, 'The Montford Delusion'.

22 'What do the "Climategate" hacked CRU emails tell us?', Skeptical Science, undated: http://bit.ly/2IkmBnC.

23 See George Monbiot's remarks on this: 'The "climategate" inquiry at last vindicates Phil Jones – and so must I', *The Guardian*, 7 July 2010: http://bit.ly/39mS8kN.

24 Rosemary Randall and Paul Hoggett, 'Engaging with Climate Change: Comparing the Cultures of Science and Activism', in Hoggett (ed.) *Climate Psychology*, Palgrave Macmillan, London, 2019, pp. 239–61.

25 Andrew Freedman, 'Science historian reacts to hacked climate e-mails', *The Washington Post*, 23 November 2009: https://wapo.st/2Tx7goU.

26 Patrick Galey, 'Climate Correction: When scientists get it wrong', Phys. Org, 23 November 2018: http://bit.ly/3aqKkhR.

27 Nicholas Lewis, 'A major problem with the Resplandy et al. ocean heat uptake paper', Climate Etc., 6 November 2018: http://bit.ly/2IllV1j.

28 Dana Nuccitelli, 'Here's what happens when you try to replicate climate contrarian papers', *The Guardian*, 25 August 2015: http://bit.ly/3cwXegj.

29 Lucy Magnan, 'Climategate: Science of a Scandal review – the hack that cursed our planet', *The Guardian*, 14 November 2019: http://bit.ly/2VITBoH.

30 Desmond Butler and Juliet Eilperin, 'The anti-Greta: A conservative

think tank takes on the global phenomenon', *The Washington Post*, 24 February 2020: https://wapo.st/3dFoEhp/.

31   'Communicating Climate Science', Science and Technology Committee, House of Commons, 2014: http://bit.ly/3aqKXIf.
32   Leo Hickman, 'BBC issues internal guidance on how to report climate change', Carbon Brief, 7 September 2018: http://bit.ly/2vFBKwV.
33   'Our Planet Matters: What's the BBC Plan all About?' BBC News, 16 January 2020: https://bbc.in/2PN73wF.
34   My redoubtable editor at Birlinn, James Rose, informs me that Carl Jung in *The Integration of the Personality* attributes the same proverb to Switzerland. I first heard it from somewhere attributed to Persia, and as it will be apocryphal, and as I've previously written of it as such, I'll risk my academic reputation and keep it that way.
35   Stanley Cohen, *States of Denial: Knowing About Atrocities and Suffering*, Polity, Cambridge, 2001, p. 58.
36   Robert Tollemache, 'We Have to Talk About . . . Climate Change', in Hoggett (ed.), *Climate Psychology*, pp. 217–38.
37   George Marshall, *Don't Even Think About It: Why Our Brains Are Wired to Ignore Climate Change*, Bloomsbury, London, 2014, pp. 227–9.
38   I have discussed Festinger's book, *When Prophecy Fails*, in more detail with Matt Carmichael in *Spiritual Activism: Leadership as Service*, Green Books, Cambridge, 2016, pp. 105–9.
39   Carl Jung, *Memories, Dreams, Reflections*, Fontana, Glasgow, 1967, p. 190.
40   Jim Morrison, 'Riders on the Storm', 1971: http://bit.ly/2xd1SQt.
41   Akwesasne Notes (ed.), *Basic Call to Consciousness*, Native Voices, Tennessee, 1978.

Chapter 6 – Rebellion and Leadership in Climate Movements

1   Greta Thunberg, 'Prove Me Wrong', Davos Speech, 22 January 2019, *No One Is Too Small to Make a Difference*, Penguin, London, 2019, pp. 19–24.
2   Greta Thunberg, Twitter, 23 September 2019: http://bit.ly/2TnKbqo.
3   Greta Thunberg, 'You're Acting Like Spoiled, Irresponsible Children', ESEC speech in *No One . . .* , pp. 34–40.
4   Greta Thunberg, Twitter, 26 December 2018: http://bit.ly/39lVdS8.
5   Greta Thunberg, 'Almost Everything is Black and White', Declaration of Rebellion XR inaugural speech, London, 31 October 2018, in *No One . . .*, pp. 6–13.
6   'Secretary General's remarks on Climate Change', UN, 10 September 2018: http://bit.ly/2TnKd12.
7   Tom Wall, 'Stroud, the gentle Cotswold town that spawned a radical protest', *The Observer*, 20 April 2019: http://bit.ly/2TnyVJX.
8   Tim Jackson, 'Zero Carbon Sooner – The case for an early zero carbon

target for the UK', CUSP working paper 18, 18 July 2019: http://bit.
ly/2IikSiL.

9   'Our Demands', Extinction Rebellion, 2018: http://bit.ly/32OkJNn.

10 Extinction Rebellion, *This Is Not a Drill: An Extinction Rebellion
    Handbook*, Penguin, London, 2019.

11 'Our principles and values', Extinction Rebellion, 2018: http://bit.
    ly/38p3xiG.

12 Damien Gayle, 'Police ban Extinction Rebellion protests from
    whole of London', *The Guardian*, 14 October 2019: http://bit.ly/2Tp
    DXWC.

13 Tom Wilson and Richard Walton, *Extremism Rebellion: A review
    of ideology and tactics*, Policy Exchange, July 2019, p. 5: http://bit.
    ly/2TmU9HZ.

14 Mark Townsend, 'Tube protest was a mistake, admit leading Extinction
    Rebellion members', *The Observer*, 20 October 2019: http://bit.
    ly/2TqMboJ.

15 Leonardo Boff and Clodovis Boff, *Introducing Liberation Theology*,
    Burns & Oates, London, 1987.

16 Starhawk, *Webs of Power: Notes from the Global Uprising*, New Society,
    BC, 2002. *The Empowerment Manual: a Guide for Collaborative Groups*,
    New Society, BC, 2011.

17 Extinction Rebellion, *This Is Not a Drill*, pp. 172–5.

18 The same issue also galvanises in discussions around the need for 'war-
    time measures' or 'emergency powers'. See Dougald Hine and Duncan
    McLaren, 'Climate emergency: the democracy fork', Open Democracy,
    11 December 2019: http://bit.ly/2wwKKEX.

19 Thank you, all those years ago, Professor Alesia Maltz of Colebrook,
    CT.

20 This drew on principles of participatory rural appraisal developed by
    Fr John Roughan of the Solomon Islands Development Trust that (as a
    member of the group) he had brought back from the South Pacific. See
    Kenyon Wright, et al., *People and Parliament: Reshaping Scotland? The
    People Speak*, 1999: http://bit.ly/2TpEIPs.

21 'Citizens' Assembly of Scotland', Scottish Government, 2020: http://
    bit.ly/39ujllG

22 'Democracy Matters', Citizens' Assembly UK: http://bit.ly/2TnMmd6.

23 'The Extinction Rebellion Guide to Citizens' Assemblies', Extinction
    Rebellion, 2019: http://bit.ly/2wrQlMz.

24 Jonathan Rose, 'The Ontario Citizens' Assembly on Electoral Reform',
    Canadian Parliamentary Review, August 2007, pp. 9–16: http://bit.
    ly/39nFKkz.

25 Isabelle Stadelmann-Steffen and Clau Dermont, 'How exclusive is
    assembly democracy? Citizens' assembly and ballot participation com-
    pared', *Swiss Political Science Review*, 22:1, 2016, pp. 95–122: http://bit.
    ly/38o6Jem.

26 Three speeches that I gave, the latter explicitly on the platform of being

a 'critical friend', can be found linked on the website of *The Ecologist*, 22 July 2019: http://bit.ly/2POmxkd.

27 'The Extinction Rebellion Guide to Citizens' Assemblies', p. 14.

28 Climate Assembly UK, House of Commons, 2020: http://bit. ly/33iRBho.

29 Karl Mathiesen, 'Leading climate lawyer arrested after gluing herself to Shell headquarters', Climate Home News, 16 April 2019: http://bit. ly/3bohIMy.

30 Zoe Blackler, 'Defence statement by Sir David King in support of five Extinction Rebellion defendants', Extinction Rebellion, 31 January 2020: http://bit.ly/2IQFCy5.

31 Roger Hallam, *Common Sense for the 21st Century*, PDF version 0.3: http://bit.ly/2PIUnqE.

32 George Monbiot, Twitter, 1 October 2019: http://bit.ly/38pkskW.

33 D. D. Raphael, *Problems of Political Philosophy*, Praeger, London, 1971.

34 'Nuke Treason Bid is Blocked,' *Sunday Mail*, 7 October 2001, p. 37: http://bit.ly/2uWJcUc. I was immediately found 'not guilty', probably as I turned up at Helensburgh Sheriff Court with a large box of legal textbooks planning to mount a defence based on what I splendidly designated, 'constitutional theology'.

35 Angie Zelter, *Trident on Trial: the Case for People's Disarmament*, Luath Press, 2001.

36 Walter Wink, *Engaging the Powers*, Fortress Press, MN, 1992.

37 Etzel Cardeña, Steven Lynn and Stanley Krippner, *Varieties of Anomalous Experience: Examining the Scientific Evidence*, American Psychological Association, Washington, 2000.

38 William James, *The Varieties of Religious Experience: a Study in Human Nature*, Fontana, Glasgow, 1960.

39 Harris Friedman and Glen Hartelius (eds), *The Wiley-Blackwell Handbook of Transpersonal Psychology*, Wiley-Blackwell, London, 2013.

40 Stanislav Grof and Christina Grof (eds), *Spiritual Emergency: When Personal Transformation Becomes a Crisis*, Tarcher/Putnam, NY, 1989.

41 A good anthology of mainly spiritual approaches to nonviolence and one that contains Gene Sharp's essay 'Disregarded History: The Power of Nonviolent Action' (pp. 231–5) that expresses a critical stance towards the spiritual, is Walter Wink (ed.), *Peace is the Way: Writings on Nonviolence from the Fellowship of Reconciliation*, Orbis, Maryknoll, NY, 2000.

42 Nafeez Ahmed, 'The flawed social science behind Extinction Rebellion's change strategy', Insurgeintelligence, 28 October 2019: http://bit. ly/2woUXDb.

43 Roberto Foa and Andrew Klassen, 'Where People are Satisfied with Democracy and Why', The Conversation, 11 February 2020: http://bit. ly/2VKY2Iw.

44 'XR Talks Non-Violent Direct Action', 3 March 2019, YouTube: http://

bit.ly/2PJyuaz. I have not watched most of Hallam's many other videos and therefore confine my comments to these two key sources.

45   Mahatma Gandhi in Thomas Merton, *Gandhi on Non-Violence*, New Directions, NY, 1965, p. 24.

46   Bible Hub, 'Praus – Strongs Greek 4239', *HELPS Word Studies*: http://bit.ly/2vH9ss1.

47   John 14:6.

48   Gandhi cited in Michael Nagler, *The Nonviolence Handbook*, Berrett-Koehler, San Francisco, 2014, p. 10.

49   Merton, *Gandhi on Non-Violence,* p. 30.

50   Xu paper used by Hallam: Yangyang Xu and Veerabhadran Ramanathan, 'Well below 2°C: Mitigation strategies for avoiding dangerous to catastrophic climate changes', *PNAS*, 114:39, 2017, pp. 10315–23: http://bit.ly/2vDnKE0.

51   BBC News, Roger Hallam interviewed by Stephen Sackur, BBC *HardTalk*, 17 August 2019: http://bit.ly/32TRdG6.

52   Scott Johnson (ed.), 'Prediction by Extinction Rebellion's Roger Hallam that climate change will kill 6 billion people by 2100 is unsupported', Climate Feedback, 22 August 2019: http://bit.ly/3atu8ww.

53   Decca Aitkenhead, 'James Lovelock: "Enjoy life while you can: in 20 years global warming will hit the fan"', *The Guardian*, 1 March 2008, http://bit.ly/39pIOwf.

54   James Lovelock's MSNBC interview. I have used this reproduction as the original is now gone: http://bit.ly/2InnhZa.

55   For example, Sam Courtney-Guy, 'Climate scientists blast Extinction Rebellion speaker who told kids they may not grow up', *The Metro*, 29 October 2019: http://bit.ly/39pMZIx.

56   Tamsin Edwards, Twitter, 26 October 2019, thread down incl. Read's reply: http://bit.ly/39nJHpp.

57   Rupert and I had an hour-long telephone conversation around these points on 31 March 2020. Our subsequent Zoom dialogue took place with over 70 participants and can be watched at: Laura Hope-Gill (chairing), 'Climate Science and Activism. Where Now? A Dialogue: Rupert Read and Alastair McIntosh', Laura Hope-Gill Vimeo, 18 April 2020: www.vimeo.com/409572058. One of the points that he made was the importance, if contrasting denialism with alarmism, of not lumping everyone into one camp or the other. There are nuances that run between and through us all.

58   University of Cumbria, 'Professor Jem Bendell, PhD', Institute for Leadership Sustainability, Business: http://bit.ly/3cxu2pt.

59   Jem Bendell, 'Doom and Bloom: Adapting to Collapse', Extinction Rebellion, *This Is Not a Drill*, pp. 73–7.

60   Jem Bendell, *Deep Adaptation: A Map for Navigating Climate Tragedy*, IFLAS Occasional Paper 2 (version July 2018 as edited December 2018): http://bit.ly/3cvbc2a.

61   Jem Bendell, 'A Year of Deep Adaptation', Professor Jem Bendell blog,

7 July 2019: http://bit.ly/2TmZgYH. Also source of the half-million downloads statistic. Note that the coronavirus is not (in any obvious way) caused by climate change.

62 Jem Bendell, 'A Summary of Some Climate Science in 2018', Professor Jem Bendell blog, 22 March 2018: http://bit.ly/2Tmq8I9.

63 Arctic News page linked by Bendell: Sam Carana, 'Warming Climate Warning!! Alert: Signs of Extinction', Arctic News, 3 March 2018: http://bit.ly/2IjywSD. I've also cited from pages linked thereto. A number of the writers featured in Arctic News, including John Nissen, were associated a decade ago with AMEG, the Arctic Methane Emergency Group.

64 Mann and Schmidt, Twitter thread, 22 November 2019: http://bit.ly/2IkoqxQ. Schmidt, first quote in the tweet, second in the Nafeez Ahmed *Vice* article linked by Mann to whom Schmidt was responding.

65 Jem Bendell, 'Responding to Green Positivity Critiques of Deep Adaptation', Resilience, 15 April 2019: http://bit.ly/2ToNWuY.

66 'Deep Adaptation Retreat with Jem Bendell and Katie Karr: Inner resilience for tending a sacred unravelling', Kalikalos Holistic Network, 2020: https://bit.ly/2vJci6Y. Also, with comments at the bottom around the dilemmas of flying to such a location: https://bit.ly/399lPEQ (2018) and https://bit.ly/39bXI8G (2019). Bendell tells that his life changed early in 2018, having taken a year's leave from his job as a professor and gone to live in Bali, Indonesia. There he made a film about his grief over climate change, released as: '"Grieve Play Love" short film on climate despair', Professor Jem Bendell blog, 24 March 2019: https://bit.ly/2VzucVD. In a subsequent video recording of being interviewed by a member of the polling company Ipsos MORI's Green Economy market research team on 3 April 2020, he speaks of being back in Indonesia once again, living under COVID-19 lockdown like in Britain, and through immersion in the garden, cooking, reading and playing the guitar, 'suddenly … somehow loosening the habit of consumption': Jem Bendell interviewed by Jessica Long, 'A Covid19–Climate connection? IPSOS Mori interviews Prof. Jem Bendell', *Jem Bendell YouTube*, April 14 2020, 16:10 minutes in to 18:30: https://youtu.be/KpNiGK3FlWA.

67 Jem Bendell: 'The Worst Argument to Try to Win: Response to Criticism of the Climate Science in Deep Adaptation', Professor Jem Bendell blog, 27 February 2020: https://bit.ly/33z2oUW.

68 Jack Hunter, 'The "climate doomers" preparing for society to fall apart', BBC News, 16 March 2020: https://bbc.in/3bh7Cr4.

69 Bendell, *Deep Adaptation*, with citation to Guy McPherson, 'Climate Change Summary and Update', Nature Bats Last, update 2 August 2016: http://bit.ly/2wwGW6h.

70 Rajani Kanth, 'On Imminent Human Extinction: [Guy McPherson] Interviewed by Rajani Kanth', Nature Bats Last, 12 October 2018: http://bit.ly/2wrYdxB. Also, Guy McPherson, Twitter, 25 September 2019: http://bit.ly/2Iij2ɪb.

71 Guy McPherson, 'Contemplating Suicide? Please Read This', Nature Bats Last, 8 July 2014: http://bit.ly/32R7lrM.

72 Rupert Read, 'After the IPCC report, #climatereality: A friendly critique of Jem Bendell's "Deep Adaptation" paper', Medium blog, 15 October 2019: http://bit.ly/2IlFt5u.

73 I. L. Janis and S. Feshbach, 'Effects of fear-arousing communications', *The Journal of Abnormal and Social Psychology*, 48:1, 1953, pp. 78–92.

74 Matthew Taylor and Jessica Murray, '"Overwhelming and terrifying": the rise of climate anxiety', *The Guardian*, 10 February 2020: http://bit. ly/2TCudqJ.

75 Antonius Robben and Marcelo Suárez-Orozco (eds), *Cultures Under Siege: Collective Violence and Trauma*, Cambridge University Press, Cambridge, 2000.

76 T. S. Eliot, 'The Dry Salvages', *Four Quartets*, Faber and Faber, London, 1944, lines 169–70.

77 Katharine Hayhoe, Twitter, 19 December 2019: http://bit.ly/2TqyRcF.

78 Michael Mann (on Guy McPherson), Twitter, 13 August 2019: http://bit.ly/2Ipx2pL.

79 Michael Mann, Twitter, 16 February 2019: http://bit.ly/2VJtmqX.

80 'Roger Hallam calls Holocaust "just another fuckery in human history"', *Die Zeit Online*, 20 November 2019: http://bit.ly/390W708.

81 Laura Backes and Raphael Thelen, 'We are engaged in the murder of the world's children', *Spiegel International*, 22 November 2019: http://bit.ly/2uSGlLU.

82 Julia Encke and Marabel Riesmeier, *Frankfurter Allgemeine*, 25 November 2019: http://bit.ly/2wwJv8p.

83 'Extinction Rebellion: Co-founder apologises for Holocaust remarks', BBC News, 21 November 2019: https://bbc.in/38iyOUk.

84 Roger Hallam, 'A response to the article in Die Zeit', Facebook, 21 November 2019: http://bit.ly/2To1fMz.

85 Freedom News, 'XR leak: Hallam Holocaust comments were a deliberate "provocation"', 24 November 2019: http://bit.ly/2ToQYzm.

86 XR Scotland statement on Roger Hallam, 3 December 2019: http://bit. ly/2vHg7MT.

87 George Monbiot, Twitter, 20 November 2019: http://bit.ly/32Qmolr.

88 Extinction Reality, 'XR Groups and Individuals Speak in Regards to Accountability in Roger Hallam Conflict Process', Twitter, 26 November 2019: http://bit.ly/2x2jmi7.

89 'Open Letter from XR Regenerative Cultures Circle Coordinators', Google Documents, 25 November 2019: http://bit.ly/32Qg79j.

90 'Update on process regarding Roger Hallam', Extinction Rebellion UK, 4 December 2019: http://bit.ly/2VFMIxc.

91 'Update about situations and processes involving Roger Hallam from the XR UK Transformative Conflict & Justice Systems (TCJ) team', Extinction Rebellion UK, 1 March 2020: http://bit.ly/3cLNVcr.

92 'Roger Hallam "Common Sense for the 21st Century"', The Sanctuary

for Independent Media, Troy, NY, YouTube video: http://bit.ly/2xr YhoU.

93 Rupert Read, Twitter, 29 February 2020: http://bit.ly/338DcVo with links to videos and pamphlet. Direct link to pamphlet co-authored with Marc Lopatin and Skeena Rathor, *Rushing the Emergency, Rushing the Rebellion?*: http://bit.ly/2UEa6KR (v. 27 January 2020).

94 This may not be comparing like with like, though, as Read doesn't state what question was being asked in the poll. By mid-April 2020, Extinction Rebellion had launched a £250,000 fundraiser and announced it had temporarily run out of funds.

95 Rupert Read, Q&A at 'XR's Rupert Read clarifies impact and purpose of blocking UK Supermarket Distribution Centres', Showroom Cinema, Sheffield, 29 November 2019: http://bit.ly/2v64dfe. Also, 'Rupert Read XR Pamphlet 2020', James Murray-White video: http://bit.ly/2PXTTgy.

96 See our study on community vulnerability and resilience in the Outer Hebrides – Lauren Eden and Alastair McIntosh, 'When the Ferries Fail to Sail', *Dark Mountain*, 5, 2014, pp. 140–57: http://bit.ly/2xnRSE5.

97 Rupert Read, Twitter, 12 March 2020: http://bit.ly/3cUGPSQ, and phone conversation, 31 March 2020. Also, note 57 above on our Zoom dialogue.

98 Aimee Groth, 'Extinction Rebellion is using holacracy to scale its international movement', Quartz at Work, 29 December 2019: http://bit.ly/2x5nuoQ.

99 Jo 'Joreen' Freeman, 'The Tyranny of Structurelessness', 1970: http://goo.gl/IlO799.

100 Benjamin Labaree, 'New England Town Meeting', *The American Archivist*, 25, 1962, pp. 165–72: http://bit.ly/2wN3akX.

101 Michael J. Sheeran SJ, *Beyond Majority Rule: Voteless Decisions in the Religious Society of Friends*, Philadelphia Yearly Meeting, Philadelphia, 1983.

102 Tova Green, Peter Woodrow and Fran Peavey, *Insight and Action: How to discover and support a life of integrity and commitment to change*' (see especially Appendix II, 'Quaker Resources on Clearness and Spiritual Discernment'), New Society, Gabriola Island, 1994.

103 Mark 9:35.

104 Norman Cohn, *The Pursuit of the Millennium*, Pimlico, London, 2004 (originally 1957), p. 13. Bullet points are quoted but abbreviated.

105 Gustavo Gutiérrez, 'The Task and Content of Liberation Theology', in Christopher Rowland (ed.), *The Cambridge Companion to Liberation Theology*, Cambridge University Press, Cambridge, 2007, pp. 19–38.

106 I sometimes use the definite article with *satya* in a way that is not usually done in translated Eastern texts. This is because I interpret *satya* as being mystically equivalent to Christianity's 'the truth'. Not all would condone this interpretation, though I am certain that Gandhi would have concurred.

107 See Chapter 6, 'Cults and Charisma', in my *Spiritual Activism*, with
     Matt Carmichael. Hermann Hess explores this in 1913 in his racking
     but arousing spiritual teaching story, 'The Poet' in *Strange News from
     Another Star*, Penguin, London, 1976, pp. 30–7: https://bit.ly/2URI510.
108 M. Hafiz Sayed (ed.), *Thus Spake ... Prophet Muhammad*, Sri
     Ramakrishna Math, Madras, 1972, p. 56.
109 Carl G. Jung, *Man and His Symbols*, Picador, London, 1978, p. 84.
110 Abraham Maslow, *The Farther Reaches of Human Nature*, Penguin, NY,
     1972.
111 Chris Rose, 'Tragedy Or Scandal? Strategies Of GT, XR and the New
     Climate Movement', Three Worlds blog, 13 February 2020: http://bit.
     ly/2ItsqPq. Full paper: http://bit.ly/3cB5RpU.
112 Likewise, the debate around green Nazism. See Franz-Josef Brüggemeier,
     Marc Cioc and Thomas Zeller (eds), *How Green Were the Nazis? Nature,
     Environment, and Nation in the Third Reich*, Ohio University Press,
     2005.
113 Mike Hulme, 'I am a denier. A human extinction denier', Climate
     Home News, 4 June 2019: http://bit.ly/2ImkEXU.

Chapter 7 – To Regenerate the Earth

1   'R. Crumb's 15 Panel Short History of America', Crumb Newsletter,
     27 April 2018: http://bit.ly/2Iiy6Mr. Click thumbnail to expand picture.
2   *An Ecomodernist Manifesto*, April 2015: http://bit.ly/3cFU2iu.
3   Stewart Brand, *Whole Earth Discipline*, Atlantic Books, London, 2010,
     p. 1; cf. Psalms 82:6.
4   Brand, *Whole Earth Discipline*, pp. 26–7.
5   Christina Waggaman, Twitter, 8 January 2020: http://bit.ly/2vvDXv4.
6   Raimon Panikkar, *The Rhythm of Being*, Orbis, Maryknoll, NY, 2010,
     p. 405. I owe much, including my subtitle, to the late Panikkarji's
     thought: http://bit.ly/32XxgOx.
7   Stan Arnaud, 'Electric plane plan for islands receives UK government
     cash', *Press & Journal*, 20 November 2019: http://bit.ly/2VJnoGw.
8   'High-Speed Rail Less Green Than UK Air Route, Says Report', *Railway
     Technology*, 17 August 2009: http://bit.ly/32QTNN5.
9   Zeke Hausfather, 'Factcheck: How electric vehicles help to tackle cli-
     mate change', CarbonBrief, 13 May 2019: http://bit.ly/2IlGf2E.
10   Katie Pavid, 'We need more scarce metals and elements to reach the
     UK's greenhouse gas goals', Natural History Museum, 6 June 2019:
     http://bit.ly/38nO7LF.
11   Laura Millan Lombrana, 'Saving the Planet with Electric Cars Means
     Strangling this Desert', Bloomberg Green, 11 June 2019: https://bloom.
     bg/2PK6FPK.
12   The Climate Coalition: http://bit.ly/38nOSV1. (At the time of writing,
     the Northern Ireland Assembly was suspended.)

13 'The Common Home Plan', Common Weal, Glasgow, 2019: http://bit. ly/2IlH8Z2.

14 Murdo MacDonald, 'The significance of the Scottish generalist tradition', in Jim Crowther, Ian Martin and Mae Shaw (eds), *Popular Education and Social Movements in Scotland Today*, NIACE, 1999, pp. 83–91. See also: http://bit.ly/2TptCtD.

15 'Mahatma Gandhi Quotes: A Collection', Gandhi Sevagram Ashram: http://bit.ly/2IpRMxF. I have rendered the quotation gender neutral.

16 See under 'Resources' at the top of https://www.ipcc.ch/srccl/.

17 Matt McGrath, 'Coronavirus forces postponement of COP26 meeting in Glasgow', BBC News, 1 April 2020: https://bbc.in/2X3r6Mo.

18 See my *Soil and Soul: People versus Corporate Power*, Aurum Press, London, 2001.

19 Michael Scott and Sarah Johnson, *The Battle for Roineabhal*, Scottish Environment LINK, 2006: http://bit.ly/3avVLVJ.

20 Alastair McIntosh and Michel Picard, 'Who is Your Enemy? Lafarge, NGOs and the Harris Superquarry Campaign', in Kai Hockerts and Luk Van Wassenhove (eds), *It's All Our Business: Corporate Responsibility in a Global World*, INSEAD Business School, Paris, Chapter 2.3: http://bit.ly/3atvgjD.

21 Larry Fink, 'A Fundamental Reshaping of Finance', BlackRock, 2020: http://bit.ly/39nam5r.

22 In their 2018 environmental report, LafargeHolcim claim a CO_2 reduction of only a quarter. This is because the figures now consolidate the erstwhile Holcim's less efficient plants.

23 Simon Dietz, et al., *Management Quality and Carbon Performance of cement producers: update*, Transition Pathway Initiative, Grantham Institute, London, 2018: https://bit.ly/343PuP6.

24 Timothy Gorringe, *The World Made Otherwise*, Cascade, Eugene, OR, 2018, p. 236.

25 Alice Walker, 'We Alone Can Devalue Gold', *Horses Make a Landscape Look More Beautiful*, Women's Press, London, 1985, p. 12.

26 John Gray, 'This Changes Everything: Capitalism vs the Climate review – Naomi Klein's powerful and urgent polemic', *The Guardian*, 22 September 2014: http://bit.ly/2uTujC5. I raise in depth the same historical points, going back to ancient Greek and Middle Eastern times, in *Hell and High Water*.

27 See my discussion of this breakdown in empathy, what T. S. Eliot called a 'dissociation of sensibility', in *Hell and High Water*, chapters 6 and 7; also in passages on trauma in *Poacher's Pilgrimage*, Chapter 12.

28 Adrienne Rich, 'Origins of History and Consciousness', *The Dream of a Common Language*, W. W. Norton, NY, 1978, p. 7.

29 Paulo Freire, *Pedagogy of the Oppressed*, Penguin, Harmondsworth, 1972, p. 21.

30 Alfredo López de Romaña (see next paragraph) used the term 'autonomous economy', of which he saw the 'vernacular' or commodity-free

economy as a subset ('The Autonomous Economy', *Interculture*, XXII:3, Part 1, p. 26, note 2: https://bit.ly/2vQPMfr). Today, however, the term 'autonomous economy' has tellingly come to mean a 'smart' or IT-driven economy, so I choose to widen Romaña's sense of the vernacular.

31 Ivan Illich, 'Vernacular Values', *CoEvolution Quarterly*, Summer 1980, pp. 1–49: http://bit.ly/3ao7xkE.

32 David Cayley, *The Rivers North of the Future: the Testament of Ivan Illich*, Anansi, Toronto, 2005, pp. 132–8.

33 Alfredo López de Romaña, 'The Autonomous Economy', *Interculture*, XXII:3–4, Part 2, pp. 1–169, 1989; quoted p. 164: http://bit.ly/3apPpHe.

34 Ugo Bardi, 'Cassandra's Curse: How the "Limits to Growth" was Demonized', Resilience, 15 September 2011: http://bit.ly/3amlhwm.

35 Julian Simon, *The Ultimate Resource*, Princeton University Press, Princeton, NJ, 1981.

36 'Reducing emissions in Scotland: 2019 Progress Report to Parliament', Committee on Climate Change, Edinburgh, 2019: http://bit.ly/2Twb4GM.

37 UN Environnent Programme, *Emissions Gap Report 2019*, Nairobi, 26 November 2019: http://bit.ly/3arVmDz.

38 Centre for Alternative Technology, *Zero Carbon Britain: Rising to the Climate Emergency*, 2019: http://bit.ly/3artK1l.

39 Future Earth and The Earth League, *10 New Insights in Climate Science in 2018*, Future Earth, 2018, p. 8: http://bit.ly/3arwICZ.

40 T. S. Eliot, 'East Coker', *Four Quartets*, Faber and Faber, London, 1944, lines 101–20.

41 Gregg Marland, Tom Oda and Thomas A. Boden, 'Cut emissions per capita to 1955 levels', *Nature*, 31 January 2019, p. 567: https://go.nature.com/3auSfKW.

42 Kate Proctor, 'Calls for Tory aide to be sacked over "enforced contraception" remarks', *The Guardian*, 16 February 2020: http://bit.ly/2VFViMq.

43 'Total Fertility Rate 2020', World Population Review, 18 October 2019: http://bit.ly/3attxLg.

44 Moldova, Pew-Templeton Global Religious Futures: http://bit.ly/2vvMPRo.

45 Madalin Necsutu, 'Falling Birth Rate Threatens Moldova's Future', Balkan Insight, 24 January 2018: http://bit.ly/2Im8kqm.

46 Jenna Vehviläinen, 'Does it make sense to pay people to have kids?', BBC Generation Project, 22 October 2019: https://bbc.in/3cv1eO5.

47 EMS Project Group, 'Who's a Real Scot: Report of Embracing Multicultural Scotland', Centre for Human Ecology, 2000: http://bit.ly/2uS5mXy.

48 Alastair McIntosh, 'St Andrew's Day "State of the Nation" Lecture', St Giles' Cathedral, Edinburgh, 30 November 2018: http://bit.ly/2IgcujL.

Chapter 8 – The Survival of Being

1   Bernard Narokobi, *The Melanesian Way*, Institute of Papua New Guinea Studies, Boroko, 1983: http://bit.ly/2IgkMrQ.

2   Michael Somare, *Sana*, Niugini Press, Port Moresby, 1975. As I've lost my copy, I'm grateful for this profile of Somare that quotes key passages: Ben Bohane, 'Somare's Choice: fight leader or peacemaker', *The Australian*, 6 February 2012: http://bit.ly/2ImbiuZ.

3   'Prof Alastair McIntosh Bntu Perencanaan Pembangunan 100 Tahun ke Depan' (Papua Province Provincial Secretary Meeting), *Cenderawasia Pos*, 9 February 2013: http://bit.ly/39q39BH.

4   Photograph of Alex Rumaseb with his books in Alastair McIntosh, Vérène Nicolas and Sibongile Pradhan, 'Healthy Community, Healthy Land: rediscovering the art of community self-governance', Report on the 2019 Papua–Scotland Study Tour, Centre for Human Ecology, Glasgow, 2019: http://bit.ly/Papua-Scot-2019.

5   See *Soil and Soul*, on the land reform campaign; and on Eigg Electric, countless YouTube videos or http://bit.ly/2TlRM8o.

6   From my friend Ralph Metzner of the Leary-Alpert-Metzner counterculture trio, at the International Transpersonal Association Conference, Killarney, Ireland, 25 May 1994.

7   Oral tradition, 'Leac nan Gillean, Suardail', Hebridean Connections: http://bit.ly/39jGWVX.

8   Anon., 'South Sea Labour Traffic', *Evening News*, Sydney, 16 April 1885, p. 4: http://bit.ly/38rS7dL.

9   Angus Macleod, 'Overview of the Pairc Clearances', Angus Macleod archive, undated: http://bit.ly/32PiqJT. Some of the houses at Croigearraidh had been rebuilt but later tumbled down again.

10 Iain Mackinnon, 'Colonialism and the Highland Clearances', *Northern Scotland*, 8:1, 2017, pp. 22–48: http://bit.ly/2TlNdLi.

11 Donald Macdonald, *Lewis: A History of the Island*, Gordon Wright Publishing, Edinburgh, 1978, p. 164: http://bit.ly/38pDXtE.

12 Steven Brocklehurst, 'Donald Trump's mother: From a Scottish island to New York's elite', BBC Scotland News, 19 January 2017: https://bbc.in/2TogXai.

13 I have explored cultural trauma at length in my other books, especially *Soil and Soul*, *Hell and High Water*, and, with nods to Donald Trump, in *Poacher's Pilgrimage*. For articles, maps and other visual supporting documentation about Trump and the Budhanais clearances, see my index page: http://bit.ly/2wswo8w.

14 J. D. Vance, *Hillbilly Elegy*, William Collins, London, 2016, pp. 1–9, my italics. WASPS = White Anglo-Saxon Protestants.

15 Michael Newton, 'Race, Whiteness and the Myth of Celtic Appalachia', *Ashville Wordfest* (videoed lecture), 2018: http://bit.ly/2wq1Zrs.

16 Sandra Dick, 'Meet Donald Trump's mother: new documentary lifts lid on Mary Anne MacLeod', *The Herald*, 14 September 2019: http://bit.ly/2TlNzBC.

17   Justin A. Frank, *Trump on the Couch*, Avery, NY, 2018, p. 142.
18   With a nod to the folk collections of Alexander Carmichael, see Alastair
      McIntosh, *Island Spirituality*, Islands Book Trust, Kershader, 2013:
      http://bit.ly/Island-Spirituality.
19   *Stornoway Gazette*, 26 January 2017, p. 1.
20   Yolanda Gampel, 'Reflections on the prevalence of the uncanny in social
      violence', in Robben and Suárez-Orozco (eds), *Cultures under Siege:
      Collective Violence and Trauma*, pp. 48–69.
21   I explore the theology and political implications of such trauma in con-
      texts wider than just Donald Trump in *Poacher's Pilgrimage*.
22   Jedediah Britton-Purdy, 'A Shared Place: Wendell Berry's Lifelong
      Dissent', *The Nation*, 9 September 2019: http://bit.ly/38tNAYE.
23   Anthony Giddens, *The Consequences of Modernity*, Polity Press,
      Cambridge, 1991, pp. 21–29: http://bit.ly/2TmP825.
24   Audre Lorde, 'Uses of the Erotic', *Sister Outsider*, Crossing Press,
      Freedom, CA, 1984, pp. 53–9.
25   See Chapter 7, 'Colonised by Death', in *Hell and High Water*, exploring
      consumerism in depth.
26   'Associated Advertising Clubs of the World', *TIME*, 25 May 1925.
27   Edward Bernays, *Propaganda*, Ig Publishing, NY, 2004.
28   Vance Packard, *The Hidden Persuaders*, Penguin, London, 1960,
      pp. 193–5.
29   Anon., 'Dichter, Ernest (1907–1991) (obituary)', AdAge Encyclopedia,
      15 September 2003: http://bit.ly/32QtB4S.
30   Ernest Dichter, 'Why Do We Smoke Cigarettes', *The Psychology of
      Everyday Living*, Barnes & Noble Inc., NY, 1947, pp. 86–99.
31   John Kenneth Galbraith, *The Affluent Society*, Marner Books, NY, 1998,
      p. 129.
32   Madonna, 'Material Girl', 1984: http://bit.ly/39k6vpQ.
33   Genesis 2:9; Revelation 2:7, 22:2; Acts 3:21. Such mystical theology is
      well developed in the Orthodox spirituality of the Eastern Church, see
      John Chryssavgis and Bruce Foltz, *Toward an Ecology of Transfiguration:
      Orthodox Christian Perspectives on Environment, Nature and Creation*,
      Fordham University Press, NY, 2013.
34   Rowan Williams, 'Afterword', *This Is Not a Drill*, Extinction Rebellion,
      London, 2019, pp. 181–4.
35   Jeremiah 2:13.
36   Micah White, *The End of Protest: A New Playbook for Revolution*, Alfred
      Knopf, Toronto, 2016, pp. 37, 241–3.
37   Anna Pigott, 'Extinction Rebellion's "regenerative culture" could be just
      as revolutionary as its demands', Open Democracy, 2 May 2019: http://
      bit.ly/32RXxos.
38   Sent by Anna Levin from an album, *Sound*, by her friend Pádraig
      Stevens; reproduced with his kind permission.
39   James Hunter, *From the Low Tide of the Sea to the Highest Mountain*

Tops: Community Ownership of Land in the Highlands and Islands of Scotland, Islands Book Trust, Kershader, 2012.

40 Alastair McIntosh, 'A "collector's item" or community ownership?' Isle of Eigg Trust launch address, *Edinburgh Review*, 88, 1992, 158–62: http://bit.ly/2wtqP9T.

41 See Community Land Scotland: http://bit.ly/32OmcmE.

42 WHFP, 'Repopulation success for community landowners', *West Highland Free Press*, 20 March 2020, p. 9.

43 'Helping Hands Across Harris, Isle of Harris Distillery', 27 March 2020: https://bit.ly/3bus2gi.

44 'Hebrides: H2. Pabbay / Pabaigh (Harris)', The Papar Project, 2005: https://bit.ly/2UuT8hF.

45 Audre Lorde, 'Uses of the Erotic', *Sister Outsider: Essays and Speeches*, Crossing Press, Toronto, 1984, pp. 53–9.

46 Interview: Martin Smith, 'What Makes a Community Resilient', *Stanford Business*, 5 April 2018: https://stanford.io/2WXyus3. Full study: Hayagreeva Rao and Henrich R. Greve, 'Disasters and Community Resilience: Spanish Flu and the Formation of Retail Cooperatives in Norway', *Academy of Management Journal*, 61:1, pp. 5–25: https://bit.ly/2WRnoVo.

47 IPCC, *AR5 Climate Change 2014: Mitigation of Climate Change*, Working Group III Contribution, Intergovernmental Panel on Climate Change, Geneva, 2014, p. xiii: http://bit.ly/3dFqxho.

Chapter 9 – The Rainmakers

1 Camilla Hodgson, 'Hottest decade ever recorded "driven by man-made climate change"', *Financial Times*, 15 January 2020: https://on.ft.com/2IkkPCW.

2 Bishop Hill, '1970s Global Cooling Alarmism', Bishop Hill blog, 1 March 2013 – thanks, yer Grace ;) – http://bit.ly/2TEcEqo.

3 'IPCC Author Nominations', IPCC: http://bit.ly/399rOKz.

4 Emiliano Rodríguez Mega, 'Clouds' cooling effect could vanish in a warmer world', Nature News, 25 February 2020: https://go.nature.com/3dEcRD2.

5 See *Hell and High Water*, Chapter 5.

6 I explore this engagement with the military in *Poacher's Pilgrimage*.

7 GalGael Trust, Twitter, 14 February 2020: http://bit.ly/2vDUSeS. Scottish Government, *National Performance Framework*, 'What is it?' and 'Sustainable Development Goals' tabs, 4 July 2018 as updated 2019 (my italics): https://bit.ly/3cP5PtY.

8 'Visit to the Western Isles', *Stornoway Gazette*, 9 April 2015, p. 13: https://bit.ly/3awH7O9.

9 Joanna Macy and Molly Brown, *Coming Back to Life: Practices to Reconnect Our Lives, Our World*, New Society, Canada, 1998.

10   Dougald Hine, 'Notes from Underground #10: The Costs of Knowing',
      Bella Caledonia, 11 February 2020: http://bit.ly/2VLUbuK.

11   Michael Mann, *Climate Change: The Science and Global Impact*, edX
      SDGacademy: http://bit.ly/2PJhRfi.

12   On some of the semantic ranges of *dhē* and *dher* see Calvert Watkins
      (ed.), *The American Heritage Dictionary of Indo-European Roots*,
      Houghton Mifflin Harcourt, Boston, 2011, pp. 18–19. Also online:
      http://bit.ly/39zE8UZ.

13   Tim Clarkson, 'Posts tagged "Doomster Hill"', Heart of the Kingdom:
      Early Medieval Govan: http://bit.ly/2wsHwSQ.

14   William Shakespeare, 'Richard the Third', IV:iv, 229, *The Complete
      Works of William Shakespeare*, vol. 8, Harap, London, undated, p. 116:
      https://bit.ly/3bGb1zM.

15   Nirmal C. Sinha, 'The Missing Context of Chos', University of
      Cambridge Repository, undated (1960s), p. 25: http://bit.ly/2PL63sW.
      This text, and discourse with Swami Sharvananda Giri and Swami
      Brahmananda of the Sivananda Ashram, Nassau, contributes to the
      sense of *dharma* that I have attempted to share here.

16   Psalms 23:4.

17   Maxim Gorky, *The Lower Depths*, Yale University Press, New Haven,
      CT, 1959, p. 18.

18   Alastair McIntosh, 'Obituary – Tom Forsyth, crofter and pioneer of
      Eigg land reform', *The Herald*, 31 August 2018: http://bit.ly/2Vth6uA.

19   Richard Wilhelm, 'Preface', *The I Ching or Book of Changes*, Routledge
      & Kegan Paul, London, 1968, pp. xlv–xlvi.

20   Meredith Sabini, *The Earth Has a Soul: C. G. Jung on Nature, Technology
      & Modern Life*, North Atlantic Books, Berkeley, CA, 2008, pp. 211–14.
      She gives two versions. I have riffed off them.

21   'Australia fires: Angry residents berate PM Morrison in blaze-ravaged
      town', BBC News, 2 January 2020: https://bbc.in/2uUAE07.

22   Carl Jung, *The Undiscovered Self*, Routledge Classics, London, 2002
      (1958), pp. 1, 16, 73–8, my italics: http://bit.ly/2T2q1BE.

23   Max Bell, 'The Doors: the story of Strange Days and the madness of Jim
      Morrison', *Classic Rock*, 12 November 2016: http://bit.ly/2Pun9Lo.

24   Freire, *Pedagogy of the Oppressed*, p. 21.

25   As befits a spiritual teaching story I've riffed this from memory of a
      recorded talk heard years ago. There's a more sedate version involving
      a lizard rather than a mouse in Ram Dass, *Changing Lenses: Essential
      Teaching Stories from Ram Dass*, Love Serve Remember Foundation, Los
      Angeles, 2018, p. 42.

INDEX